Metallophytes in Mineral Mining

METALLOPHYTES IN MINERAL MINING

Vimal Chandra Pandey
Centre for Research Impact and Outcome, Chitkara University Institute of Engineering and Technology at Chitkara University, Rajpura, Punjab, India

Dragana Ranđelović
Institute for Technology of Nuclear and Other Mineral Raw Materials, Belgrade, Serbia

Ksenija Jakovljević
Institute for Biological Research "Siniša Stankovic", National Institute of the Republic of Serbia, University of Belgrade, Belgrade, Serbia

Ritu Chaturvedi
Department of Botany, St. John's College, Agra, Uttar Pradesh, India

ELSEVIER

Elsevier
Radarweg 29, PO Box 211, 1000 AE Amsterdam, Netherlands
125 London Wall, London EC2Y 5AS, United Kingdom
50 Hampshire Street, 5th Floor, Cambridge, MA 02139, United States

ISBN: 978-0-443-30110-0

For Information on all Elsevier publications
visit our website at https://www.elsevier.com/books-and-journals

Publisher: Peter Llewellyn
Acquisitions Editor: Jennette McClain
Editorial Project Manager: Stuti Das
Production Project Manager: Sathyavani Deiva Sigamani
Cover Designer: Mark Rogers

Typeset by MPS Limited, Chennai, India

Contents

About the authors

Dr. Vimal Chandra Pandey is an applied research scientist internationally recognized for his extensive work in phytomanagement of polluted sites. He is consistently ranked among the world's top 2% of scientists (2020–2025), according to Stanford University and Elsevier BV. His pioneering research focuses on the remediation and management of heavy metal–polluted sites, especially fly ash deposits, coal mine sites, and bauxite residue (red mud) using ecologically and socioeconomically valuable plants to restore ecosystem services and support a bio-based economy. His interests also include advancing phytoremediation to utilize polluted lands and contribute to achieving the United Nations Sustainable Development Goals. His work has expanded the scope of phytoremediation beyond its traditional applications. He is currently engaged in exploring commercial phytoremediation approaches that involve minimal risk, low input, and low maintenance. Dr. Pandey is currently associated with the Centre for Research Impact and Outcome, Chitkara University Institute of Engineering and Technology at Chitkara University, Rajpura, Punjab, India. He is also the president of the "Society for Green Development," Lucknow, Uttar Pradesh, India. Prior to his current position, he worked at the National Botanical Research Institute,Babasaheb Bhimrao Ambedkar University, and the Council of Science and Technology, Lucknow, Uttar Pradesh, India. Dr. Pandey is the recipient of several awards, honors, and fellowships. He serves as a review expert and a member of international research evaluation commissions and panels. He has more than 150 scientific publications to his name. He is also the author and editor of several influential books published by leading international publishers, with more forthcoming. Dr. Pandey serves as an associate editor, academic editor, and editorial board member for several prestigious international journals with significant impact factors. He has also served as a guest editor for many reputed international journals.

Dr. Dragana Ranđelović works as a senior research associate at the Institute for Technology of Nuclear and Other Mineral Raw Materials, Belgrade, Serbia. She holds a BS and an MS from the Faculty of Forestry in Landscape Architecture and a PhD in Multidisciplinary Sciences for Environmental Protection from the University of Belgrade, Belgrade, Serbia. She is working on the reclamation and remediation of degraded lands using selected plants, with a focus on metal and metalloid transport and accumulation in the soil–plant system. She investigates these processes in a variety of mine wastes and other anthropogenically altered environments, such as ruderal habitats or regions affected by environmental accidents. Additionally, she examines different categories of plant taxa occurring in degraded areas—sensitive and endangered, medicinal, invasive, or ruderal species—determining their potential for phytoremediation. She has published more than 60 international and national scientific papers dealing with various environmental topics.

Dr. Ksenija Jakovljević is a plant ecologist specializing in the study of plant–soil relationships on metalliferous sites, currently working as a principal research fellow at the Institute for Biological Research "Siniša Stanković"—National Institute of Republic of Serbia, University of Belgrade, Belgrade, Serbia.

She holds a BS, an MS, and a PhD from the same faculty. She is mainly involved with metallophytes, both from soils naturally enriched with metals and metalloids, such as ultramafic soils, and from anthropogenically contaminated substrates, focusing on the study of soil–plant systems and the discovery of new hyperaccumulator species and their potential application for phytoremediation. Her research includes diverse plant groups: endemic, medicinal, ruderal, and invasive plant species. She has published more than 50 international and national scientific papers on various aspects of plant research, including flora, vegetation, biogeography, conservation, and plant ecology.

Dr. Ritu Chaturvedi is an environmental botanist engaged in research on plant responses to heavy metal stress. Her research encompasses processes from plant physiology, microbial ecology, and environmental remediation. She did her PhD from the Department of Botany, St. John's College, Agra, Uttar Pradesh, India. Her early research was focused on bioremediation of heavy metals through fungi and ethylenediaminetetraacetic acid. She is also working on plant-based bioremediation or phytoremediation. Till now, she has six research papers, seven book chapters and one book entitled "Eco-Restoration of Mine Land". She also serves as a potential reviewer for several reputed national and international journals.

Preface

Continuous increase in the number and area of metal- and metalloid-rich soils is a major concern worldwide. Contaminated areas denote a major challenge across the world for the survival of living organisms, including humans. Several areas and ecosystems have been converted into major toxic hotspots, and many are likely to be added to the list. Metal-rich soils and mainly ultramafic soils (i.e., serpentine soils) are widespread all over the world and could be used for metal mining by metallophytes. Metallophytes that accumulate extraordinary amounts of metals or metalloids within their living tissues are called hyperaccumulators. Phytomining or agromining involves the cultivation of hyperaccumulator plants and harvesting their biomass to obtain particular metals or metalloids. Whereas phytomining describes the process of exploiting plants for valuable elements, agromining refers to the full agronomic chain in using accumulator/hyperaccumulator plants as "metal crops." The process involves the farming of "metal crops" on industrial or mineral wastes in order to obtain valuable element(s) from their harvest biomass via the production of a "bio-ore." The demand for critical metals, including platinum group elements (PGEs), rare earth elements (REEs), cobalt, nickel, and even gold, is more under pressure in the 21st century as a result of resource depletion and geopolitical factors.

This book presents the current state of knowledge on the global distribution and ecology of hyperaccumulator plants, the biogeochemical pathways involved in these processes, the influence of rhizospheric microbes, as well as aspects of propagation and conservation of these unusual plants. This book then summarizes the state of the art in using new tools for hyperaccumulator discovery and in the understanding of their physiology and molecular biology. The opportunities for incorporating agromining into rehabilitation and mine closure strategies are presented, and ecosystem service provision and life cycle analysis are discussed. The possibilities for agromining nickel, cobalt, manganese, selenium, arsenic, thallium, REEs, and PGEs are discussed in separate chapters devoted to case studies. Lastly, an overview concludes the book, looking ahead to prospects for the development of phytomining in the future.

This book is authored by pioneers in the field who have been at the forefront of the phytomanagement of polluted sites over the past 15 years. The book is timely, as phytomining is now a key idea in metal farming development, with the rapid expansion of mining activities worldwide. Thus, the present book will be of significant interest to environmental professionals in the minerals industry, policy makers, stakeholders, government regulators, practitioners, and academia.

Vimal Chandra Pandey
Dragana Ranđelović
Ksenija Jakovljević
Ritu Chaturvedi

Acknowledgments

We sincerely express our gratitude to Jennette McClain (Senior Acquisitions Editor) and Stuti Das (Editorial Project Manager) from Elsevier for their excellent support, guidance, and coordination during the production of this fascinating project. Finally, a massive thank you to our families for their endless support, patience, and encouragement.

CHAPTER ONE

Metallophytes

1.1 Introduction

Habitats that differ in their characteristics for plants often represent stress of different kinds, varying in duration and intensity. One of the most common stress factors to which plants are exposed in their habitat is stress caused by metals. Habitats that are characterized by an increased concentration of metals and metalloids in the substrate are referred to as metalliferous, regardless of whether they are naturally or anthropogenically enriched. Depending on the element present in excessive concentrations, a distinction is made between ultramafic (serpentinised) soils (rich in nickel, iron, magnesium, chromium, and cobalt), calamine soils (rich in zinc, lead, and cadmium), seleniferous soils (rich in selenium), and so on (Raskin et al., 1994).

Most plants are sensitive to such an increased content of elements in the substrate that cause some damage, or rarely resistant when tolerance mechanisms allow them to survive. Plants that are adapted to these specific soils, in which they survive and reproduce without showing symptoms of toxicity, are called metallophytes (Baker et al., 2010). Depending on whether they are exclusively associated with sites with excessive elemental concentrations or not, these plants can be obligate and facultative (pseudo-) metallophytes. Obligate metallophytes are species that are restricted to areas with a certain composition of elements in excessive concentrations (Wójcik et al., 2017), which is why they are also geobotanical indicators (Brooks, 1994). Many of these species are endemic and represent both paleo- and neo-endemic species. However, facultative metallophytes are much more numerous, as they are adapted to a wider range of substrate characteristics. They are also referred to as pseudometallophytes or bodenvag species, as they are not strictly tied to a specific substrate type, although these terms are more applicable to indifferent species (Pollard et al., 2014). They are mostly ecotypes that have evolved due to metal stress and are tolerant to excessive element concentrations in the soil. *Noccaea caerulescens* (J.Presl & C.Presl) F.K.Mey. and *Arabidopsis halleri* (L.) O'Kane & Al-Shehbaz are the

Metallophytes in Mineral Mining
DOI: https://doi.org/10.1016/B978-0-443-30110-0.00004-0

most studied pseudometallophytes, model species for the study of the phenomenon of accumulation, but also *Silene latifolia* Poir. and *Minuartia verna* (L.) Hiern (syn. *Sabulina verna* (L.) Rchb.), which are very important for their accumulation specificity (Jakovljević et al., 2024a).

1.2 Origins and development of metallophytes

Metallophytes are plants that thrive in soils with high concentrations of metals and metalloids such as nickel, lead, zinc, cadmium, arsenic, thallium, and tolerate them. Their emergence and development is closely associated to evolutionary adaptations that enable them to survive in such extreme environments. Metalliferous sites induce two basic responses in their flora: plants either avoid these substrates or colonize them with varying degrees of success. Plants that are able to colonize metalliferous soils, so-called metallophytes, can be (1) indifferent to these specific substrate types, (2) physiologically tolerant specific edaphic race or ecotypes or even and (3) taxonomically recognized endemic species if speciation become strong enough. As they are thought to have evolved through parallel evolution in different types of habitats, ecotypes are usually referred to as neoendemics, while palaeoendemics are species that once had a large distribution range but are now restricted to much smaller and localized areas and are now independent, well-separated taxa. Metal toxicity as a strong selective factor can cause evolutionary change in just one generation, assuming that plants have sufficient genetic variability to support this (Bradshaw, 1984). In this sense, the evolution of taxa on ultramafics is particularly interesting, as it is a substrate type that supports a specific type of flora and represents an important refugium and center of speciation, as in the central Balkan Peninsula (Jakovljević et al., 2011).

Several hypotheses have been proposed to explain the role of excessive elemental concentrations in plants (Boyd, 2014; Rascio and Navari-Izzo, 2011): (1) metal disposal; (2) interference hypothesis; (3) drought resistance; and (4) the elemental defense hypothesis. Metal disposal was found to be extremely important for plants living on metalliferous sites and unable to avoid elemental uptake. The concentrations of Ni and Cr, for example, can be extremely high in ultramafic sites, and these elements, although essential, are only required by plants in small quantities. The excess of elements are deposited in metabolically less active or inactive sites to protect plant physiology. These are mostly epidermal tissues (at the tissue level) and vacuoles and cell walls at the cellular level (Rascio and Navari-Izzo, 2011). The interference hypothesis states that hyperaccumulator plants prevent the establishment of less tolerant plants through metal-rich litter. This was shown with *Stanleya pinnata* (Pursh) Britton and *Astragalus bisulcatus* (Hook.) A.Gray, Se-hyperaccumulators that enriched the soil with Se and prevented the germination of Se-intolerant species (El Mehdawi et al., 2011a). However, the same authors showed that positive effects on plant growth were also observed (El Mehdawi et al., 2011b). In view of the fact that some species do not create a toxic environment due to the low availability of elements, this theory is not yet widely accepted. Drought stress is often observed in metalliferous sites, especially those with ultramafic rocks, which is mainly caused by the hostile physical properties of the overlying soils and the open character of the vegetation (Kruckeberg, 2002). This theory is based on the assumption that metals act as osmolytes to retain water in the cell walls; however, this theory is yet to be confirmed. The defense hypothesis, which is based on the role of metal accumulation against herbivores and pathogens, is the most intensively researched (Boyd, 2014). This theory has been confirmed in several studies in both hyperaccumulators and accumulators and also in the context of organic plant defenses (Behmer et al., 2005; Boyd, 2007; Jhee et al., 2006), but not for many species and many different elements (Ni was analyzed in half of the studies (Boyd, 2007), and for the most part not yet confirmed in the field). Although hyperaccumulation has been experimentally confirmed as a defensive trait, the opposite has been found for some insects that feed exclusively on plants rich in certain elements. These include *Melanotrichus boydi* Schwarz & Wall, which specializes on *Streptanthus polygaloides* A.Gray and *Chrysolina pardalina* Fabricius, which exclusively feeds on the leaves of *Berkheya coddii* Roessler (Boyd, 2014).

Although tolerance to excessive elemental concentrations is thought to have evolved independently and there is no link to specific plant families, representatives of the Brassicaceae, Euphorbiaceae, and Asteraceae families appear to be most commonly found in metalliferous sites (Cappa and Pilon-Smits, 2014). This is particularly true for hyperaccumulator plant species, with representatives of the Brassicaceae family being the most common (with 83 species out of a total of 721 [Reeves et al., 2018]). Although hyperaccumulation of Zn (*Noccaea* and *Cardamine*), Pb (*Noccaea*), Cd (*Noccaea*), and Tl (*Iberis* and *Biscutella*) has been confirmed, most species hyperaccumulate Ni (predominantly representatives of the genera *Noccaea* and *Odontarrhena*) (Jakovljević et al., 2024a; LaCoste et al., 1999; Reeves et al., 2018).

In some species, the phenomenon of multitolerance has been observed, that is plant species developed tolerance to more than one element present in excessive concentrations. Such examples of adaptation should be sought on substrates such as ultramafics, rich in several trace elements (Ni, Cr, Co, and Fe) and on mineralized soils rich in Zn, Pb, and Cd or Tl and As. Multitolerance has been found in several *Noccaea* species, such as *N. caerulescens* and *Noccaea praecox* (Wulfen) F.K.Mey, which are tolerant to excessive concentrations of Ni, Zn, Pb, and Cd (Krämer, 2010; Likar et al., 2010), then in *M. verna* (Johnston and Proctor, 1977; Jakovljević et al., 2024a) (tolerant to high concentrations of Pb, Zn, As, and Tl; a), *Viola arsenica* Beck, *Viola allchariensis* Beck, and *Viola tricolor* subsp. *macedonica* (Boiss. & Heldr.) A.F.W.Schmidt (As and Tl; Bačeva Andonovska et al., 2021). Although tolerance to several different elements is a relatively common phenomenon, hyperaccumulation of several elements has been observed in a small number of species, especially for three elements. Simultaneous hyperaccumulation of four different elements is not yet known in nature (Gei et al., 2020).

The best-known metallophytes occur in ultramafics, geologically unique areas rich in metals and metalloids, and are called ultramaficophytes. The group of environmental constraints that ultramafics induce in plants as a stressful environment has become known as the "serpentine syndrome" (Jenny, 1980). This syndrome leads to a number of morphological adaptations in plants, such as nanism, stenophylly, plagiotropism, glaucescency, and glabrescence (Kruckeberg, 1992). The main challenges of ultramafic soils include (1) a low Ca/Mg ratio in the soil; (2) nutrient deficiencies; and (3) excessive concentrations of elements characteristic of ultramafic soils, such as Ni, Fe, Mg, Cr, and Co (Kazakou et al., 2008). A low Ca/Mg ratio has been emphasized as one of the main causes of plant fertility, as it has been shown that a high Ca/Mg ratio can mitigate stress effects, even when multiple limitation is present (Galey et al., 2017). The antagonistic relationship between Mg and Ca resulted in the suppression of Ca by excessive Mg concentrations and localized root tip necrosis (O'Dell and Claassen, 2006). However, the necrosis observed in nonultramafic, but not in ultramafic plants, suggests physiological mechanisms that enable the plant to tolerate and cope with high Mg concentrations relative to low Ca. There are several hypothesized mechanisms in plants under low Ca/Mg stress: more efficient uptake of Ca than of Mg, especially compared to nonultramafic plants. In addition, Ca is transported more efficiently to the leaves, while Mg remains largely bound in the roots (O'Dell et al., 2006).

However, a high demand for Mg has also been proposed as one of the hypotheses, which is due to a limited uptake of Mg (Kruckeberg, 1954). In addition to the low Ca/Mg ratio, soil fertility is severely impaired by nutrient deficiencies, particularly a lack of N, P, and K. As the geographical relationship of soils deficient in certain elements has shown (N is largely absent in the USA and P in European soils), it is most likely that deficiencies of several elements are responsible for stress in plants and not just one particular element (Chiarucci et al., 1999; O'Dell and Claassen, 2006; Nagy and Proctor, 1997). Excessive concentrations of elements characteristic of ultramafic soils (Ni, Fe, Mg, Co, and Cr) represent the additional stress that plants have to cope with. The main focus is on Ni enrichment, as high Fe and Mg concentrations are common in plants. Concentrations of Co are not as excessive as Ni, while high Cr concentrations in soil are never taken up by plants due to its extremely low availability (Kazakou et al., 2008). Although they have adapted well to Ni-rich soils, there is no evidence that ultramaficophytes require higher Ni concentrations for their growth (Reeves and Baker, 1984).

1.3 Categories of metallophytes

Excessive elemental concentrations in the substrate induce three types of responses in metallophytes: exclusion, indication, and accumulation (Baker, 1981).

- **Excluders** are plants that keep the concentration of elements in their aerial parts constant and low despite excessive concentrations in the soil, which are either not absorbed or are absorbed at the root level, preventing their transport to the aerial organs. If the metal(oid) concentrations in the soil exceed the critical value, the uptake of elements becomes unrestricted.
- **Indicators** are plants in which the concentration of elements in the shoot is regulated by the concentration in the substrate.
- **Accumulators** are plants that concentrate metals and metalloids in the shoots over the both low and high concentrations in the substrate.

The exclusion of metals and metalloids proved to be the most frequently observed tendency in nature (Bothe, 2011). There are two mechanisms for the exclusion of elements, either avoiding uptake into the root or uptake into the roots but limiting translocation into the aerial tissues (root accumulators). Since root uptake is difficult to avoid, limiting translocation to the aerial tissues is the most common control mechanism (Baker, 1981).

In plants that have developed mechanisms of detoxification of elements in aboveground organs, significant concentrations of elements are translocated to the shoots, with the leaves being the most common site of deposition of metals and metalloids. These plants are considered accumulators. At the extreme end of this group are the hyperaccumulators, that is, plants in which the concentrations of certain elements are 2–3 orders of magnitude higher than the concentrations in the leaves of nonaccumulators. These values are recognized as threshold values for hyperaccumulation and are set at different levels for the various elements. It is 100 $\mu g\ g^{-1}$ for Cd, Se, and Tl, 300 $\mu g\ g^{-1}$ for Co, Cr, and Cu, 1000 $\mu g\ g^{-1}$ for As, Ni, and Pb, 3000 $\mu g\ g^{-1}$ for Zn, and 10,000 $\mu g\ g^{-1}$ for Mn (van der Ent et al., 2013). To be considered a hyperaccumulator, plants must meet other criteria in addition to exceeding the threshold for hyperaccumulation, such as a bioconcentration factor >1 (more often >50), a shoot-to-root ratio >1, and effective detoxification (Baker and Whiting, 2002). Only plants that fulfill these criteria in the natural population can be recognized as hyperaccumulators, but not experimentally cultivated plants.

Due to intensive research in recent decades and the application of new methods (e.g. X-ray fluorescence spectroscopy [XRF]), the number of hyperaccumulator species has constantly changed, with the most recently determined number being 721 taxa (Reeves et al., 2018). Most of these species are hyperaccumulators of Ni (523 taxa), whereas hyperaccumulation of other elements was found in a much smaller number of species. For example, hyperaccumulation of copper was found in 53 taxa, hyperaccumulation of cobalt and magnesium in 42 each, and selenium in 41. Foliar concentrations above the hyperaccumulation threshold for Zn were found in 20 species (with 31 taxa recognized by Vaculík et al. (2025)), lead in 8 taxa, cadmium in 7 taxa, arsenic in 5, rare earth elements and thallium in 2, and chromium in 1 (Reeves et al., 2018).

Hyperaccumulation of Ni is the most widespread in nature, and occurs in more than 70% of the hyperaccumulator species known to date (Reeves et al., 2018). Such a large number of nickel hyperaccumulators is the result of the distribution of ultramafics, the main source of Ni in nature, which are both refugia and centers of flora speciation (Jakovljević et al., 2011). The use of dimethylglyoxime (DMG), a rapid method mainly used in the field for detecting high Ni concentrations in leaves (Reeves et al., 1996), also contributed to the discovery of a considerable number of hyperaccumulators of this element. Hyperaccumulation of Ni was most frequently found in *Odontarrhena* and *Noccaea* species, both belonging to the Brassicaceae family. Some of these species are considered hypernickelophores, as defined by Jaffré et al. (2013), with >1 wt % Ni in the leaves, such as in *Odontarrhena lesbiaca* Candargy with 23,650 $\mu g\ g^{-1}$ (Kazakou et al., 2010), Odontarrhena *baldaccii* (Vierh. ex Nyár.) Španiel with 17,670 $\mu g\ g^{-1}$ Ni (Reeves et al., 1997), *Noccaea boeotica* F.K.Mey. with 23,400 $\mu g\ g^{-1}$ Ni (Reeves and Brooks, 1983), but also in *Bornmuellera* species -Bornmuellera *baldaccii* (Degen) Heywood, *Bornmuellera dieckii* Degen, *Bornmuellera emarginata* (Boiss.) Rešetnik (Fig. 1.1A), and Bornmuellera *tymphaea* (Hausskn.) Hausskn., all with more than 2 wt% Ni in the leaves (Jakovljević et al., 2022; Paçarizi et al., 2020; Reeves et al., 1983). In addition to these two genera, the most extreme Ni concentrations were found in the foliar tissues of *B. coddii* (up to 3.6 wt%; Rue et al., 2020) and in the species of the genus *Phyllanthus* (Phyllanthus *rufuschaneyi* Welzen, R.W.Bouman & Ent and Phyllanthus*balgooyi* Petra Hoffm. & A.J.M.Baker; van der Ent et al., 2015), *Rinorea* (1.78 wt% Ni in Rinorea *bengalensis* (Wall.) Kuntze; Reeves, 2003), *Centaurea* (1.14 wt% Ni in Centaurea *thracica* (Janka) Gugler (Fig. 1.1B); Jakovljević et al., 2022), while up to 25 wt% was found in

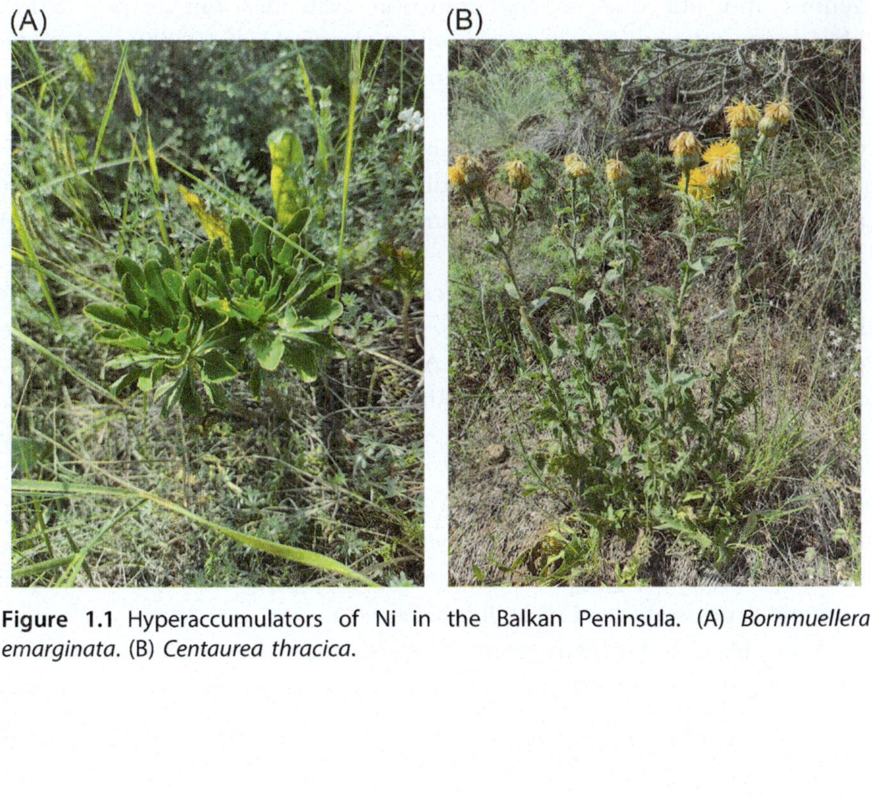

Figure 1.1 Hyperaccumulators of Ni in the Balkan Peninsula. (A) *Bornmuellera emarginata*. (B) *Centaurea thracica*.

the latex of *Pycnandra acuminata* (Pierre ex Baill.) Swenson & Munzinger (Jaffré et al., 1976).

Hyperaccumulation of copper was found in 53 taxa (Reeves et al., 2018). The threshold for hyperaccumulation, previously set at 1000 $\mu g\ g^{-1}$ (Baker and Brooks, 1989), was eventually corrected and lowered to 300 $\mu g\ g^{-1}$ (van der Ent et al., 2013). Most of these species originate from the Democratic Republic of Congo, from the Central African Copperbelt, where the first plants tolerant to extremely high concentrations of copper and cobalt were described (van der Ent et al., 2019). In some of these species, even the previous threshold of 1000 $\mu g\ g^{-1}$ was exceeded (Blissett, 1966). However, the easy contamination of the plant material by soil and dust and their difficult removal, especially in species with a dense trichome layer, raise doubts as to whether contamination is the cause of such high concentrations (van der Ent et al., 2019; Faucon et al., 2007). The same applies to Co, which was found in the leaves of 42 taxa in concentrations above the hyperaccumulation threshold (Reeves et al., 2018), although in some of them (such as *Haumaniastrum katangense* (S.Moore) P.A.Duvign. & Plancke) SEM-EDS and synchrotron X-ray absorption spectroscopy revealed that the hyperaccumulation was endogenous and not due to contamination (van der Ent et al., 2019). Hyperaccumulation of manganese with a threshold of 10,000 $\mu g\ g^{-1}$ (van der Ent et al., 2013) was also found in 42 taxa (Reeves et al., 2018), which, in addition to plant affinity, depend more on the pH of the substrate than on the pseudo-total and extractable concentrations of this element (Fernando et al., 2008). The distribution of these species was mainly restricted to New Caledonia, Australia, China, and Japan (Reeves et al., 2021).

As it is one of the essential microelements that must be continually biofortified, the discovery of new hyperaccumulators of selenium has proven to be extremely important. Among the 41 known Se hyperaccumulators, the most intensively studied taxa are *A. bisulcatus*, *Neptunia amplexicaulis* Domin and *S. pinnata*, with *A. bisulcatus* having the highest concentrations found under natural conditions (ca. 14,000 $\mu g\ g^{-1}$; Sura-de Jong et al., 2015). Although Se concentrations were highest in the young leaves of all three species (Harvey et al., 2020), different patterns of Se translocation and deposition indicate that hyperaccumulation developed multiple times (van der Ent et al., 2019).

Zinc, lead, and cadmium are elements that often occur together in nature, both in mineralized areas and in mine waste. The soil particularly rich in these elements is called calamine and are inhabited by a considerable number of locally distributed taxa or their ecotypes, such as *Viola lutea* subsp. *calaminaria* (Ging.) Nauenb. and *Viola lutea* var. *westfalica* A.A.H.Schulzand *N. caerulescens* subsp. *calaminaris* (Lej.) Holub (Wójcik et al., 2017). Hyperaccumulation of Zn, an element essential for plant growth, was found in 20 taxa (Reeves et al., 2018), in which foliar concentrations of this element exceeded the threshold of 3000 $\mu g\ g^{-1}$ (van der Ent et al., 2013), which is reduced from the previously proposed 10,000 $\mu g\ g^{-1}$ (Baker and Brooks, 1989). Although some of these plants have been found on Zn-poor soils with concentrations in the range of 10–100 $\mu g\ g^{-1}$ (Dinh et al., 2015; Jakovljević et al., 2023a; Stein et al., 2017; Vaculík et al., 2025), most of these plants live near Zn mines and smelters with extremely high concentrations of this element in the substrate (Reeves and Baker, 2000). Hyperaccumulation of Pb and Cd is found in a much smaller number of taxa (eight and seven, respectively; Reeves et al., 2018). Although extremely high concentrations were found especially near mining and smelting areas, the bioavailable concentrations of Pb were considerably lower, mainly due to complexation with organic matter and clay present in the soil (Tang et al., 2009), but also due to low mobility through plant tissue (Yang et al., 2021). Hyperaccumulative Pb concentrations in foliar tissues were therefore much more often due to uncontrolled uptake from the extremely contaminated substrate (e.g. >10,000 $\mu g\ g^{-1}$ Pb), or contamination and insufficient washing, than to the affinity of plants to take up elements. Hyperaccumulation of Cd is rather rare, mainly due to weak accumulation. However, at Cd-contaminated sites, which are also not so common, even in mine waste areas, foliar concentrations can be >100 $\mu g\ g^{-1}$, even >1000 $\mu g\ g^{-1}$, well above the threshold for hyperaccumulation (van der Ent et al., 2013), but still with large variations between samples.

Arsenic and thallium are among the least frequently hyperaccumulated elements, detected in only five and two taxa, respectively (Reeves et al., 2018). Arsenic contamination, both of natural and anthropogenic origin, poses an increasing threat to the environment. Mining and smelting of As-rich ore is one of the main sources, although the contribution from industry and pesticide use has also been shown to be significant (Robinson et al., 2006). In nature, it normally occurs as arsenite As(III) and arsenate As (V), which differ in their solubility. The more toxic and mobile As(III) is usually abundant in a more reducing environment at a pH of <8, whereas As(V) enters the cell via phosphate transporters due to their chemical similarity, disturbing the balance in P uptake (Souri et al., 2017). The fern *Pteris vittata* L. is, with 11,000 μg g^{-1}, the best-known hyperaccumulator of this element (Wei et al., 2020), although hyperaccumulation has also been shown in other fern species such as *Pityrogramma calomelanos* (L.) Link (>6000 μg g^{-1}; Visoottiviseth et al., 2002). In addition to ferns, As hyperaccumulation was detected in leaf samples of four plants from the Allchar mine in North Macedonia (*Alyssum kavadarcense* Micevski, *M. verna*, *Plantago lanceolata* L., and *Silene vulgaris* (Moench) Garcke) and in the seeds of *V. allchariensis* (Fig. 1.2), a steno-endemic species from the same area (Bačeva Andonovska et al., 2021). The highest As concentrations found in the foliar tissues of these species were in the range of

Figure 1.2 *Viola allcharensis*, the hyperaccumulator of thallium.

1160–2870 μg g^{-1} (Jakovljević et al., 2024a). Concentrations of thallium, a nonessential and highly toxic element, do not normally exceed 1 μg g^{-1} in the Earth's crust (Belzile and Chen, 2015), but there are areas that are extremely rich in this element, such as North Macedonia, France, the United States, Switzerland, and China (Liu et al., 2019). Although hyperaccumulation of thallium has only been found in two taxa, according to Reeves et al. (2018), its concentration above the hyperaccumulation threshold (100 μg g^{-1}; van der Ent et al., 2013) has shown to be more common. The highest concentrations to date have been found in leaves of *S. latifolia* Poir at 79,200 μg g^{-1} (Jakovljević et al., 2025), *V. arsenica* and *V. allchariensis* at 58,900 and 23,100 μg g^{-1}, respectively (Jakovljević et al., 2023a), and *Biscutella laevigata* L. at 32,700 μg g^{-1} (Corzo Remigio et al., 2022), but also in *M. verna*, *S. vulgaris*, *Iberis linifolia* subsp. *intermedia* (Guers.) Kerguélen and the endemic Allchar species *A. kavadarcense*, *Centaurea leucomalla* Bornm. and *Knautia caroli-rechingeri* Micevski (Bačeva Andonovska et al., 2021). As both As and Tl are enriched in sulfide minerals, their joint enrichment has also been observed in nature. High concentrations of these two elements were found in the Allchar region, the abandoned mine site on the slopes of the Kožuf Mountain in North Macedonia, a globally unique site with extremely high concentrations of As and Tl in the soil (142 g kg^{-1} As and 17.6 g kg^{-1} Tl; Đorđević et al., 2021). Co-accumulation of these two elements was found in three species from this area (*M. verna*, *P. lanceolata*, and *S. vulgaris*; Jakovljević et al., 2023a, 2024b), with the species favoring Tl more strongly. Previously, this phenomenon was only known from *P. vittata*, with >11,100 μg g^{-1} As and 111 μg g^{-1} Tl (Wei et al., 2020).

The increased demand for rare earth elements (REEs) in recent decades has stimulated research into potential hyperaccumulators of these elements. This group of elements consists of 15 lanthanides with yttrium and scandium, which are categorized into light rare earth elements (LREEs) and heavy rare earth elements (HREEs) mainly based on their electron configuration (Khan et al., 2017). Most of the hyperaccumulators of REEs belong to the pteridophytes, with >1000 μg g^{-1} of REEs in the fronds. These include *Blechnum orientale* L., *Carya glabra* (Mill.) Sweet, Carya *tomentosa* (Lam.) Nutt., and *Dicranopteris linearis* (Burm.f.) Underw. (van der Ent et al., 2021). Hyperaccumulation was also detected in flowering plants, such as *Phytolacca americana* L. (hyperaccumulator of REE and Mn) with 1040 μg g^{-1} REEs in the leaves, and more than in the stem and roots (Liu et al., 2021), but also in *Juglans nigra* L., which showed hyperaccumulation of yttrium, which is used as a proxy for REEs (Goudard et al., 2024). A series of new studies on REE hyperaccumulation will contribute not only to the knowledge on the tolerance of plants to this group of elements but also to the definition of the new threshold for hyperaccumulation, which seems to be too high and should probably be lowered to 100 $\mu g\,g^{-1}$ (van der Ent et al., 2021).

Most of the hyperaccumulator species take up only one element above the nominal threshold. Only a few species showed the ability to accumulate two elements, with simultaneous hyperaccumulation of Ni and Zn being the most frequently observed. Analysis of elemental concentrations in herbarium specimens of the genus *Noccaea* deposited in the herbarium of the National Museum of France using XRF revealed co-accumulation of Ni and Zn in 10 specimens belonging mainly to *N. caerulescens* (subsp. *caerulescens*, *brachypetala* (Lej.) Holub and *virens* (Jord.) Kerguélen), as well as in one specimen of *Noccaea violascens* (Schott & Kotschy) F.K.Mey., whereas most specimens showed hyperaccumulation of only one of the analyzed elements (Jakovljević et al., 2024a). Field studies on *N. caerulescens* also showed hyperaccumulation of Cd, and the same was true for *N. praecox* and *A. halleri* (Krämer, 2010), with even some evidence of hyperaccumulation of Pb. The hypertolerance for these three co-occurring elements could involve the same loci or these plants could have evolved on nonmetalliferous soils and later colonized metalliferous soils where their tolerance increased (Krämer, 2010). The simultaneous hyperaccumulation of Tl and As was also confirmed in plants, which, in addition to a *P. vittata*, a known hyperaccumulator of As (Wei et al., 2020), was also confirmed in four specimens in the Allchar region, most of them with predominant affinity for Tl accumulation (Jakovljević et al., 2023a, 2024b). To date, hyperaccumulation of four elements has not been confirmed in neither one species (Gei et al., 2020).

Taxonomically, most hyperaccumulator species belong to the Brassicaceae and Phyllanthaceae family (83 and 59 species, respectively) and are mainly distributed in Cuba (128 taxa), New Caledonia (99 taxa), and Turkey (59 taxa) (Reeves et al., 2018, 2021). As most Ni hyperaccumulators are associated with ultramafic areas, most Ni hyperaccumulators have been detected in New Caledonia, South Africa, Cuba, Brazil, Japan, California, Queensland, Turkey, and the Balkan Peninsula, in areas that can be broadly categorized into two zones: Mediterranean and tropical and subtropical areas around the world (Reeves et al., 2021). In the Mediterranean region, the most common Ni hyperaccumulators are from the Brassicacae family, whereas in the latter group, those from the Phyllanthaceae, Violaceae, Salicaceae, Euphorbiaceae, and Rubiaceae families predominate (Reeves et al., 2021).

The elemental accumulation is a strategy that is less frequently observed in nature and is generally more common in obligate metallophytes, which are strictly limited to metalliferous soils. However, it is not an unknown phenomenon in nonmetalliferous soils either. In fact, in a number of species, accumulation is a constitutive feature that occurs on both metalliferous and nonmetalliferous sites. One example of this is *Cardamine waldsteinii* Dyer (fam. Brassicaceae), the only species of this genus with a proven ability to accumulate Zn (Fig. 1.3; Jakovljević et al., 2023b). High foliar concentrations, above the threshold for hyperaccumulation of Zn (3000 $\mu g\ g^{-1}$; van der Ent et al., 2013), were detected by XRF in herbarium specimens in BEO (Thiers, 2024). Analysis of field samples confirmed this high affinity for Zn accumulation, with 2210 $mg\ kg^{-1}$ found in the leaves of *C. waldsteinii*, which is rather exceptional considering that only 12.7 $\mu g\ g^{-1}$ Zn was found in soil (Jakovljević et al., 2023b). More striking results were found in two subspecies of *Dichapetalum gelonioides* (Roxb.) Engl., *D. gelonioides* subsp. *sumatranum* Leenh. and *D. gelonioides* subsp. *pilosum* Leenh. growing on nonmetalliferous soil at 26,400 and 15,700 $\mu g\ g^{-1}$ Zn in the leaves, respectively (Baker et al., 1992). In *A. halleri*, which grows on Zn-poor

Figure 1.3 *Cardamine waldsteinii*, hyperaccumulator of Zn.

soils, even more extreme concentrations of 53,900 $\mu g\ g^{-1}$ were observed in the leaves (Stein et al., 2017). Hyperaccumulation of Zn as a constitutive trait was also found in *N. caerulescens* (Seregin et al., 2021), where Zn reached 9000 $\mu g\ g^{-1}$ in the leaves of plants growing on overburden with only 139 $\mu g\ g^{-1}$ Zn (Reeves et al., 2001). In this species, the expression of genes from the ZnT family was found to be higher than in the related excluder species *Thlaspi arvense* L., where their expression was observed exclusively under Zn deficiency, suggesting that hyperaccumulation in *N. caerulescens* is a constitutive response to Zn deficiency (Seregin et al., 2021). Hyperaccumulation as a constitutive trait has also been found in other elements, such as Se, which was found in the leaves at concentrations of 4000 $\mu g\ g^{-1}$ in *N. amplexicaulis* growing in soil with $<10\ \mu g\ g^{-1}$ of this element (Galeas et al., 2007).

1.4 Metal tolerance mechanisms in metallophytes

Plants have different requirements for the elements in nature and, depending on this, elements can be essential or nonessential, respectively. Elements such as Zn, Ni, Mo, Cu, Mn, and Cr are essential for the growth and development of plants and lead to stress if they are present in insufficient concentrations. The opposite is true for nonessential elements (such as Pb, As, Cd, and Tl), which can cause plant stress even in low concentrations. There are also elements that are not essential but due to their important role in plant biology, at least for some plant species, are considered beneficial, and such is the case with Na, Co, Al, and Si. Although plants need them in varying amounts, what all elements have in common is that they cause plant stress when they are present in higher concentrations than necessary. Some plants develop mechanisms to cope with these unfavorable environmental conditions, known as metal tolerance. In this sense, metallophytes are particularly interesting as they grow successfully in habitats that are lethal to most other species without showing symptoms of toxicity (Antonovics et al., 1971).

Depending on how the plants react to excessive elemental concentrations, one can speak of plant sensitivity and resistance (Baker, 1987). While sensitivity refers to the plant in which metal stress leads to plant impairment and eventual death, resistance is the plant's response that allows it to thrive, develop, and reproduce in this stressful environment. There are two strategies to achieve resistance: avoidance and tolerance. Tolerance is particularly interesting because plants can maintain a normal level of physiological processes in the presence of an excess of metals or metalloids. Tolerance occur primarily in obligate and facultative metallophytes, while accidental metallophytes can be ignored, as they are primarily associated with nonmetalliferous sites (Baker, 1987). Although the increased tolerance to metals is genetically determined, the costs of metal tolerance (slower growth and lower biomass production) should not be neglected (Ernst, 1976).

Physiologically, metal tolerance can be categorized either as exclusion, where metals are retained in the roots and their transport to the shoot is restricted, or as accumulation, where elements are preferentially transferred to the aerial parts of the plant (Baker, 1981). The main characteristics that distinguish (hyper)accumulator plant species from excluders (with a strong uptake of elements in the root) are the rapid translocation of elements to the aerial organs and sequestration in the leaves (Rascio and Navari-Izzo, 2011).

The activities of plants to eliminate metal ions from the metabolism and prevent physiological stress are both metal-specific and plant-specific. The increased uptake is made possible by the overexpression of genes such as those of the ZIP family (zinc-regulated transporter iron-regulated transporter proteins), which regulate the uptake of Zn by coding for cation transporters in the plasma membrane (Assunção et al., 2001). While the number of ZIP genes in Zn-(hyper)accumulator plant species is independent of the available Zn, an increased number of ZIP genes in nonaccumulator plants was only observed in Zn deficiency. Zinc transporters have also been shown to regulate Cd uptake, but with a preferential uptake of Zn (Zhao et al., 2002), although an independent system for Cd uptake has been confirmed in roots of *N. caerulescens* (Lombi et al., 2001). The transport system from the ZIP family is also involved in the uptake of Ni, which is translocated to the leaves after uptake into the xylem and sequestered mainly in the vacuoles as nonactive sites (Deng et al., 2018). As no high-affinity transporters for Ni are known so far, this element is mainly taken up as Ni^{2+} via low-affinity transport systems in both accumulator and nonaccumulator plant species. Uptake as Ni^{2+} is important because soluble Ni can enter the roots by coopting uptake mechanisms of cations of similar charge and size (Ca^{2+}, Cu^{2+}, Mg^{2+}, and Zn^{2+}) (Deng et al., 2018). It has been shown that Fe transporters, whose expression was stimulated by exposure to Ni, are also involved in the uptake of Ni. After penetration into the root symplast, the Ni ions are complexed to mitigate toxicity.

Increased synthesis of histidine has observed in Ni-(hyper)accumulator species, and its role in Ni chelation prevents sequestration of Ni in root vacuoles and enhances upward transport (Kozhevnikova et al., 2014). No specific efflux transporters for Ni have been identified, and leaf transpiration appears to be responsible for Ni transport after reaching the xylem vessels (Robinson et al., 2003). Most of the Ni in the xylem sap remains as a free hydrated cation (up to 70%), as shown in several *Odontarrhena* species, while only a small portion was chelated by histidine, nicotianamine, glutamine, and citrate (Krämer et al., 1996; Mari et al., 2006). After reaching the leaves, the leaf veins allowed the transport of Ni through the apoplastic region, which either remains in the apoplast or, more likely, is absorbed by the symplast. Since they are the least active, the leaf epidermal cells and their vacuoles are the most common site for Ni localization (Küpper et al., 2001). When excessive concentrations of Ni^{2+} are transported into the leaves, the apoplast also becomes very important for storage, and high Ni loading can be found in the cuticle, trichomes, cell walls, or even leaf tips, suggesting that guttation also can be a mechanism for the release of excess Ni (McNear et al., 2005; Robinson et al., 2003). Phloem translocation has been shown to be extremely important in hyperaccumulator plant species (Reeves et al., 1996; van der Ent et al., 2015), especially considering that the reproductive organs, which are the main sink for phloem translocation, are important localization sites for metals and metalloids (Bačeva Andonovska et al., 2021). In contrast to the xylem, which transports exclusively upward, transport via the phloem is bidirectional, both upward and downward, with movement to the reproductive organs and young leaves predominating (Deng et al., 2016).

1.5 Ecophysiology and ecology of metallophytes

Stress conditions on metalliferous sites cause a physiological response of varying type and intensity in plants, leading to reduced growth, mainly due to nutrient deficiency and impaired photosynthesis, and, in extreme cases, even to complete inhibition of plant growth and function (Khanna et al., 2019). The effects of metal stress on photosynthesis have been observed at many levels, including stomatal conductance, photosynthetic pigment synthesis, alteration of chloroplast structure, and so on. Changes in chloroplast ultrastructure lead to significant disturbances in photosynthesis, mainly through a reduction in the number and size of chloroplasts and changes in thylakoid organization (Barceló et al., 1988; Jin et al., 2008), although the effects are different for different metals (Ahmad et al., 2019). Changes in photosynthetic intensity caused by excessive metal(loid) concentrations in leaves include between others: the substitution of Mg^{2+} in the chlorophyll molecules by some of the divalent metal cations in the chlorophyll molecule, resulting in the inhibition of photosynthesis as normal energy transfer is not possible due to these altered chlorophyll molecules; changes in the activity of supramolecular structures such as PSI and PSII; and altering the activity of enzymes involved in the Calvin cycle, the xanthophyll cycle, and photosynthetic pigment biosynthesis (Sharma et al., 2020). A reduced gas exchange rate was observed in a number of plants under the influence of various metals and metalloids. Excessive As concentrations led to a reduction in stomatal conductance and photosynthetic rate in *Phaseolus vulgaris* L., *Avena sativa* L., and *Lactuca sativa* L. (Ahmad et al., 2019), while effects of Al on gas exchange were found in various citrus species (Pereira et al., 2000) and Cu in *Vitis vinifera* L. (Cambrollé et al., 2015).

The plant response to the stress has also been observed in anthocyanins, which are not only involved in the coloring of flowers and fruits but also play an important role in the response to excessive concentrations of potentially toxic elements, as their metal-induced accumulation has been observed (Naing and Kim, 2021). They are involved in nonenzymatic oxidative stress and prevent oxidative damage by metal chelation. Although a concomitant reaction with enzymatic components was observed, anthocyanins were found to be more important in protecting plant cells against metal stress. Not all elements induce the synthesis of anthocyanins equally. Whereas Cu stimulates the synthesis most strongly, the weakest response was found for Pb (Baek et al., 2012), provided that the metal-induced anthocyanins generally proved to be efficient only at moderate contamination but not at highly excessive element concentrations.

Numerous studies indicated that plants exposed to various stress conditions (excess metals in the substrate, drought, salt stress, high radiation) accumulate proline, which may be an important component of the nonenzymatic response of plants. Plants respond to metal stress by proline accumulation differently, depending on their level of tolerance. Whereas in nontolerant ecotypes of some well-known metallophytes such as *S. vulgaris*, metal stress induce significant accumulation of proline, in tolerant ecotypes, constitutive high concentrations of proline were found. A similar effect was found for metal-tolerant ecotypes of *Armeria maritima*

(Mill.) Willd. (Schat et al., 1997). Proline may also play a role in protecting certain proteins, such as nitrate reductase, when plants are exposed to potentially toxic elements or osmotic stress, and in stabilizing ribonucleases and proteases when plants are exposed to metal(loids) (Liang et al., 2013).

One of the symptoms of stress is also lipid peroxidation, as reactive oxygen species can cause oxidative damage to biological molecules, including lipids. Malondialdehyde (MDA), a cytotoxic product of this process is one of the reliable indicators of this process. An increase in the concentration of metals in plant tissues can lead to an increased intensity of lipid peroxidation, resulting in an accumulation of MDA (Li et al., 2013). In addition to metal stress, increased production of free radicals and accumulation of MDA have also been observed in response to salt stress and drought (Choudhary et al., 2007).

The symbiosis with arbuscular mycorrhizal fungi and the binding of metals by fungal cells is one of the mechanisms that help plants to cope with excessive elemental concentrations in the substrate. Although mycorrhiza was thought to be absent in many hyperaccumulators, it has been found to be much more abundant and has even been detected in some members of the Brassicaceae, the most numerous of the hyperaccumulators known to date, which were previously thought to be a nonmycorrhizal family (Alford et al., 2010; Leyval et al., 1997). However, mycorrhiza can not only prevent the uptake of excessive elemental concentrations but also improve nutrient uptake by increasing the underground surface area. In this respect, mycorrhiza provide plants with essential elements such as Zn, Cu, and P, which are very important for plant nutrition (Alford et al., 2010).

In addition to metal stress, metallophytes have to cope with a number of physico-chemical constrains of the substrate. These include water deficits, extreme temperature fluctuations, nutrient deficiencies, an unfavorable Ca content, and an unfavorable Ca/Mg ratio. All this makes the plants' response more complex and cannot be explained by excessive elemental concentrations alone.

1.6 The habitats of metallophytes

Metallophytes are a group of plants that can grow in sites with excessive concentrations of metals and metalloids (metalliferous sites) without showing symptoms of toxicity. These plants have developed mechanisms to tolerate such specific conditions that would be toxic or even lethal to most other plants.

Depending on the origin of the metals and metalloids, there are two main groups of metalliferous sites: those of natural and anthropogenic origin. Ultramafics are the best-known type for naturally occurring excessive elemental concentrations (Fig. 1.4). In addition to high concentrations of Fe and Mg (defined by the ferromagnesium character of the substrate), excessive concentrations of Ni, Cr, and Co are also found (Kierczak et al., 2021). Poor nutrient status and an unfavorable ratio of Ca and Mg are additional and sometimes even more important factors influencing the distribution of ultramafic metalliferous flora. Some of the most important ultramafic areas are in California and the Mediterranean region. In Europe, the largest ultramafic areas are located in the Balkan Peninsula, especially in eastern Bosnia and Herzegovina, western and eastern Serbia, Albania, and Greece (Stevanovic et al., 2003). Mineralized areas rich in other metals and metalloids, such as calamine areas rich in Pb, Cd, and Zn, or areas rich in Tl and As, are also important natural sources of

Figure 1.4 Mount Ozren, ultramafic area in southwest Serbia.

metals and metalloids. The richest calamine vegetation in Europe was found in Belgium, Poland, France, and Germany, while in Asia in Afganistan, Pakistan, Iran, and China are the most represented in this respect (Wójcik et al., 2017). Selenium-rich soils or seleniferous soils are commonly found in South Asia, Australia, Ireland, and the USA, while copper and cobalt have the highest concentrations in the so-called Copperbelt in the Democratic Republic of Congo and Zambia (El Desouky et al., 2009). Soils near volcanoes and soils affected by lava or volcanic ash are one of the main sources of metalliferous sites rich in Ni, Zn, Co, and so on, especially those prone to re-eruptions. The excessive element concentrations are closely related to the substrate characteristics in the area. For example, in the soil near the volcano in the Changbai Mountains, Zn concentrations were observed to be above background levels, while in the volcanic soil in the Solofrana River valley, Cr and Cu were above acceptable levels (Adamo et al., 2003).

Metalliferous sites with natural mineralization as the source of excessive metal concentrations, where these concentrations result from weathering, are also referred to as primary sites (Baker et al., 2010). As they show no anthropogenic influence, these sites are nowadays extremely rare and irregularly distributed all over the world, including Europe. Considering the fact that they are characterized by poor nutrient status and excessive concentrations of metals and metalloids, the productivity of these sites is very low.

However, anthropogenically contaminated metalliferous sites are much more diverse, with mine sites (including ore mining and processing) predominating, usually with extremely high elemental concentrations. There are three main types of waste generated during the extraction and processing of ores: overburden, flotation residues, and slags (Manca et al., 2021), sites with different physico-chemical characteristics, but all extremely rich in metals and metalloids. The uniqueness of these sites was already recognized in the sixteenth century, when *M. verna* was identified as a species associated with metalliferous habitats (Brooks, 1994). Industrial or domestic waste, which is usually rich in metals and metalloids, also produces metalliferous habitats that are only covered by very tolerant species. This also applies to roadsides and fallow land, which contain metals and metalloids from vehicle exhausts, industrial waste, and improper waste disposal. When affected by mining activities, metalliferous sites are considered secondary, with soils with highly altered physico-chemical properties, poor in nutrients, with low water capacity, and rich (often extreme) in metals and metalloids (Baker et al., 2010). In addition to tolerance to excessive metal concentrations, these deficiencies in nutrients and water capacity therefore represent an additional stress for these plants (Ernst, 2000).

The third group of sites are tertiary sites, which are the result of the input of excess metal into uncontaminated sites (Baumbach et al., 2007). As they are usually far away from the actual metalliferous sites and require rapid adaptation to the changed conditions, the flora in these areas is usually poor and consists of grasses with a short life cycle (Baker et al., 2010). Some of the species found here are so-called "transitional endemics," that is, species that have arrived with the ore, such as *M. verna* and *N. caerulescens*. The tertiary sites are often found in alluvial deposits created by metal pollution from rivers as riverbanks downstream of mine sites are polluted with significant amounts of metals and metalloids, leading to the occurrence of metallophytes. However, the irregular influx of potentially toxic elements, which is directly dependent on the intensity of mining activities, can affect the survival of these sites and these types of vegetation (Rodwell et al., 2007).

1.7 Metallophyte vegetation

Metallophyte vegetation refers to plant communities consisting of metallophytes—plants that are adapted to metalliferous habitats and thrive despite the hostile physico-chemical properties of these substrates. Metalliferous sites, both natural and anthropogenic, harbor a very particular vegetation characterized by three main features: (1) low productivity, (2) high endemism, and (3) distinct vegetation (Whittaker, 1954). The predominant species in metallophyte vegetation belong to the two main groups of metallophytes: excluders and accumulators (including hyper-accumulator plant species).

1.7.1 Vascular plant communities

Due to the specific and often extreme conditions, not many plant species are able to successfully colonize metalliferous sites. Two species that are particularly characteristic of metalliferous areas in Europe are *M. verna* and *S. vulgaris*, the two Caryophyllaceae species that are very tolerant of excessive concentrations of numerous metals and metalloids. Representatives of the Poaceae family are also frequently found at these sites. They form monocultures and begin to improve soil properties and the process of succession. For example, *Agrostis capillaris* L. and *A. stolonifera* Leers, *Elymus repens* (L.) Gould, and *Calamagostis epigejos* (L.) Roth are widespread in mining areas in Belgium, Poland, and the Netherlands (Wójcik et al., 2017). Species such as *C. epigejos*, which are very well adapted to disturbed habitats, form monodominant and long-lived stands in a short time, preventing further natural development of the vegetation. The initial processes of overgrowth are often associated with allochthonous and even invasive species, as these have a great potential to occupy habitats, especially those disturbed by humans. Over time, the number of species increases, although it is still lower compared to nonmetalliferous sites. Species diversity is further increased by species that have found an optimal ecological niche at these extreme sites, as well as by some microorganisms, invertebrates, and small vertebrates.

Floristic studies of metalliferous habitats have confirmed that their biodiversity is much lower compared to the surrounding areas, with differences also observed between different mine sites. Banásová et al. (2006) pointed out differences between Cu-rich and Zn–Pb-rich dumps. While lichens were found in large numbers in Cu-rich dumps, xerothermophilic plants were more common in Zn–Pb dumps. Due to their low competitiveness, the abundance of metallophytes remains stable in this vegetation, whereas they are being suppressed by other species in nonmetalliferous sites. Fertilization has a particularly strong effect on the richness of metalliferous flora and vegetation, with higher nutrient concentrations allowing competition with less tolerant plants and a reduction in the number of metallophyte species present. This has been observed in the *Violetum calaminariae* Schwickenrath community, whose representatives have a very limited distribution in western and central Europe (Willems, 2004).

Most of the sintaxa observed in metalliferous habitats in Europe are assigned to the class *Violetea calaminariae*, although a few communities fit better into other classes. Within the order *Violetalia calaminariae* Br.-Bl. et Tx. ex Ernst 1965, three regions with characteristic vegetation can be distinguished: (1) vegetation of the *Thlaspion calaminarii* Ernst 1965 alliance, from WC & C Europe, dominated by *N. caerulescens* (syn. *Thlaspi alpestre* subsp. *calaminare*) and *A. halleri*; (2) vegetation of the *Armerion halleri* Ernst 1965 alliance, characterized by the absence of *N. caerulescens* and is mainly distributed in Central Europe; (3) the metalliferous vegetation in alpine areas, which is assigned to the alliance *Galio anysophylli– Minuartia vernae* and contains representatives such as *Galium anisophyllon* Vill., *M. verna*, *Poa alpina* L., *Dianthus sylvestris* Wulfen, *Euphrasia salisburgensis* Funck ex Hoppe (Ernst, 1990).

1.7.2 Bryophyte vegetation

The presence of bryophytes (mosses and liverworts) on metalliferous sites was overlooked for a long time until they have recently received more attention due to their specific mechanisms of elemental uptake and metabolism.

The bryophyte vegetation has been studied in detail in Ireland, with 50 mine sites included in the study (Holyoak, 2008). The results indicated the presence of several distinct communities, although without any indication of a possible formal name. A number of the bryophyte species observed were categorized into different grassland and heathland syntaxa, but not clearly enough to be classified under *Violetalia calaminariae*.

The indicator species of bryophyte metallophyte vegetation indicated that they are mainly associated with calamine sites, with *Ditrichum plumbicola* Crundw. as an obligate Pb bryophyte and Ditrichum *cornubicum* Paton, *Scopelophila cataractae* (Mitt.) Broth., *Cephaloziella massalongi* (Spruce) Müll. Frib. and Cephaloziella *nicholsonii* Douin occurring exclusively on copper-bearing sites. *Weissia controversa* var. *densifolia* (Bruch, Schimp. & W.Guembel) Wilsonand *Bryum pallens* (Sw. ex Brid.) Schwägr. facultatively inhabited Zn- and Pb-rich sites, while *W. controversa* var. *densifolia*, *Scapania compacta* (Roth) Dumort., *Cephaloziella stellulifera* (Taylor ex Carrington & Pearson) Croz., and Cephaloziella *integerrima* (Lindb.) Warnst. were found as facultative accumulators on Cu-rich sites. Vascular species associated with these bryophyte species included *Plantago maritima* W.P.C. Barton and *A. maritima* (sites high in Cu, Pb and Zn), *M. verna* (sites high in Zn and Pb) and *Agrostis vulgaris* With. and *Calluna vulgaris* (L.) Hull on sites high in Cu (Holyoak and Lockhart, 2011).

1.7.3 Lichen vegetation

The ability of lichens to survive under exceptionally adverse conditions, especially as pioneer species, is well known in nature (Favas et al., 2014; Favero-Longo, 2014). Although stress related to climatic conditions has been the most studied so far, their sensitivity to other abiotic factors has become the focus of research. As their nutrition is partly related to the substrate, a response to metal stress on metalliferous sites can also be expected. Their tolerance has mainly been attributed to the fungal component, which has been intensively studied in this area under the aspect of the positive effects of mycorrhiza. Knowledge of lichens inhabiting metal-rich habitats has shifted scientific interest from the study of lichen associations to the physiological basis of their tolerance. Previous studies have shown that the number of lichen species differs considerably on different substrate types, with the greatest diversity observed on serpentinite, even greater than on other mafic rocks (Favero-Longo et al., 2004). The specific chemical composition of ultramafics and the lower competition observed in these areas is most likely the reason for this unusual abundance of lichens (Piervittori et al., 2004).

Depending on their preference for excessive concentrations of certain metals, two main alliances have been distinguished in previous studies—*Acarosporion sinopicae* (Wirth, 1972) and *Lecideion inopsis* Purvis in Purvis et Halls 1996 (Wirth, 1972). The indicator species of the Fe-associated alliance *Acarosporion sinopicae* are, besides *Acarospora sinopica* (Wahlenb.) Körb., *Lecanora gisleriana* Müll.Arg., *Rhizocarpon oederi* (Ach.) Körb., and Rhizocarpon *furfurosum* H.Magn. & Poelt, while *Psilolechia leprosa* Coppins & Purvis and *Lecidea inopis* Th.Fr. are indicator species of *Lecideion inopis*, the only alliance associated with Cu-rich habitats (Huneck, 2006).

1.8 Metallophytes: The unique biological resources of metal-rich ecosystems and their economic, ecological, and bioprospecting significance

Metallophytes are plants that can survive and develop in sites rich in metals and metalloids without showing symptoms of toxicity. In this hostile environment, of both natural and anthropogenic origin, metallophytes play a crucial role in various ecological, scientific, and economic contexts. While inhabiting metalliferous sites, these plants contribute to sustainability by either reducing elemental concentrations in the soil or preventing erosion in areas with degraded substrates. Plants that (hyper)accumulate metals that are abundant in the substrate by thriving in that substrate will reduce elemental concentrations over time. This is even more pronounced when these species are removed from the metalliferous site after reaching maturity (which is the main principle of agromining). Where accumulator species were left after completion of the life cycle, the metal cycle was observed, with high elemental concentrations found in the litter. In excluders, where most of the accumulated elemental concentrations remain in the root or the elements are not taken up at all, the prevention of erosion is seen as the main benefit. This applies to many species with an extensive root system belonging to the Poaceae family, such as *C. epigejos* (Ranđelović et al., 2018), but also to species such as *Tussilago farfara* L., from the Asteraceae family (Jakovljević et al., 2020). This is particularly important for mine tailings, where the physical properties of the substrate are very hostile and unstable. As not many plant species are able to thrive in metalliferous habitats, metallophytes contribute strongly to biodiversity with their specific tolerance, albeit with a lower species richness compared to the nonmetalliferous areas. The hostile environmental conditions have led to speciation, the intensity of which is highly dependent on the characteristics of the substrate, i.e. how extreme it is, because it is known that evolutionary changes can take place in just one generation if the plants have sufficient genetic variability to support this (Bradshaw, 1984). Both neo- and palaeoendemics have been found in this type of habitat, depending on how long they have been present.

Besides the binding of substrate particles and the extraction of excess elemental concentrations, metallophytes have an untapped potential for applications in biotechnology and the pharmaceutical industry. A number of metallophytes that can thrive on metalliferous sites are medicinal plants that can be used despite the excess of elements in the soil (either they do not take them up at all or they are not taken up in the parts that are used for medicinal purposes). This is the case with *T. farfara*, a medicinal plant often found on abandoned spoil heaps rich in As, Sb, Zn, and Pb (Jakovljević et al., 2020), which has been shown to have a positive effect on microbial substrate properties (Stefanowicz et al., 2015). *Artemisia absinthium* L., *Crataegus monogyna* Jacq., and *Hypericum perforatum* L. are medicinal plants that have also been observed in metalliferous sites. The negligible elemental concentrations in the plant tissues indicate their safe use for medicinal purposes (Badea, 2015). The metal content in the secondary metabolites of aromatic plants is also low, which is why their use can be beneficial even when they grow on highly contaminated soils, as is the case with *Matricaria*, *Ocimum*, and *Rosmarinus* species (Ranđelović, 2023).

The utilization of metallophytes also offers a number of economic advantages. The best known is phytoremediation, that is, the use of plants to remove or degrade pollutants in the environment. Depending on the mechanisms used to remove the pollutants, there are different phytoremediation techniques: phytoextraction, phytostabilization, phytovolatilization, phytofiltration, and phyto/rhizodegradation (Favas et al., 2014), with phytoextraction and phytostabilization being the most commonly used. Phytoextraction is the basis of phytomining, in which hyperaccumulating plants are used to concentrate metals in harvestable plant parts. After harvesting and drying, the metals, mainly in the form of salts, are recovered by incineration. As it involves the cultivation of hyperaccumulator plant species to extract various metals and metalloids, this technology is more accurately referred to as agromining (van der Ent et al., 2015).

1.9 Practical uses of metallophytes plants

Metallophytes play an essential role in their ecosystems by contributing to soil stabilization, providing habitats for various organisms and influencing the local nutrient cycle. They can also be used in bioremediation to remediate contaminated soils. They provide various ecological services that are essential for the sustainability of the ecosystem and the health of living organisms.

As inhabitants of substrates with unfavorable physical properties, especially mine waste sites, metallophytes play an important role in binding the soil, stabilizing it, and preventing erosion. This is especially true for species from the Poaceae family, such as *C. epigejos* (Ranđelović et al., 2018) and *Vetiveria zizanioides* (Mickovski and van Beek, 2009), which bind the substrate with their roots; the roots reduce runoff from the surface and the loss of sediments. In this way, the metallophytes prevent additional leaching of pollutants into the surrounding areas. In addition to improving the physical properties of the soil, there are also a number of benefits for the chemical composition of the substrate. These species not only reduce the bioavailable concentrations of elements through accumulation but also improve the organic matter content and, when Fabaceae species are used, also nitrogen fixation. This further increases the plant biomass of both legumes and non-legumes (Frérot et al., 2006) without increasing the use of fertilizers.

Metallophytes are involved in the biogeochemical cycling of elements in metalliferous soils, which is particularly important for elements that are present in excess. Depending on their tolerance strategy, these plants avoid the uptake and accumulation of metals and metalloids to varying degrees. This leads to a redistribution of elements, which stabilizes the chemical composition of the soil and benefits living organisms. This is particularly true for the (hyper)accumulation of elements, where the removal of metals from the soil and their binding in metabolically less active tissues leads to less contamination of the substrates over time.

As they tolerate excessive element concentrations, metallophytes contribute significantly to biodiversity. They provide ecological niches in areas that are habitable for most species and create microhabitats for other plants, animals, and microorganisms. Biodiversity is also enriched by endemic species that have evolved through the process of speciation and are found in extreme habitats such as metalliferous ones. These include the species of the genus *Viola*, which have evolved over time in habitats rich in metals and metalloids, developing not only specific ecotypes but also new subspecies or species. Closely related to the alpine *Viola lutea* are two endemic metallophytes—the yellow *V. lutea* subsp. *calaminaria* and the blue *V. lutea* subsp. *westfalica*, which are the result of speciation in extremely Zn-rich habitats (Bothe, 2011). Similarly, *V. tricolor* subsp. *macedonica*, an infraspecific taxa belonging to *V. tricolor*, showed multiple tolerance to high element concentrations, considering that it contains 4290 $\mu g\ g^{-1}$ Tl in the leaves (Bačeva Andonovska et al., 2021), but also occurs on Pb–Zn-rich mining sites in the same geographical zone. In recent decades, the conservation of metallophyte biodiversity has been one of the demands of mine site remediation, with a focus on increasing the number of species rather than the richness of individuals in monodominant stands (Baker and Whiting, 2008).

The benefits of pollination by metallophytes have also been recognized, as the flowers of some metallophytes attract pollinators, contributing not only to the success of plant reproduction but also to the population of pollinators. However, care must be taken with species whose flowers have excessive concentrations of elements, as there is a risk that metals and metalloids will be transferred to the pollinators and jeopardize their vitality. This applies to species such as *Sambucus nigra* L., *Tilia tomentosa* Moench, *Taraxacum officinale* F.H.Wigg., all of which have the potential for Cd accumulation (Bărbulescu et al., 2022).

The important role of metallophytes is also recognized in carbon storage, which is common to all plant species. In this way, plants help to mitigate climate change, which is particularly important in degraded areas such as mining and industrial areas.

In addition to these environmental benefits, the use of hyperaccumulator plant species, as the most specific group of metallophytes, has been shown to be of great economic importance for the recovery of metals and metalloids in areas where conventional mining would not bring sufficient economic gain. This method of elemental recovery using plants, or phytomining, is increasingly referred to as agromining, as it involves the cultivation of plants and the use of a range of agricultural measures.

1.10 Future prospects

In recent decades, metallophytes have received increasing scientific attention due to their physiological peculiarities and their tolerance to excessive, sometimes even extreme, elemental concentrations. Some of the trends in future research are application of metallophytes, especially some specific groups such as (hyper)accumulator plant species in phytoremediation. Although research to date has pointed to a considerable number of species that can be used in the remediation of industrial and mining waste and other sites contaminated with potentially toxic elements (PTE), further research is needed both to increase the efficiency of the (hyper)accumulators known to date and to discover the potential for the use of PTE in species previously unknown for these properties. It is particularly important to focus on the traits that most often limit the effectiveness of phytoremediation, such as the simultaneous accumulation of several different PTEs, a significant plant yield and a high biomass.

Technologies such as Clustered Regularly Interspaced Short Palindromic Repeats (CRISPR) and genetic engineering, where the plant genome is modified to improve the uptake of elements, can be used to assess the potential to improve remediation efficiency. In-depth knowledge and modifications in the overexpression of metal transporters such as those of the ZIP family as well as proteins such as metallothioneins and phytochelatins can improve metal accumulation and the overall process of detoxification. The focus should also be on microbial studies to explore possibilities to improve remediation performance by mycorrhiza and rhizobacteria.

As an environmentally friendly alternative to conventional mining, future studies on metallophytes will probably largely relate to the agromining and highlight the potential for greater efficiency. Large-scale field trials are essential for studies that have shown promise in assessing economic justification and potential environmental impact. However, considering that large metalliferous areas are located in underdeveloped countries, scaling up research and promoting sustainable practises will lead to a number of local, environmental, social, and economic challenges that will need to be overcome to achieve the expected results.

1.11 Conclusion

Metallophytes are a unique group of plants that can thrive in environments with high concentrations of metals and metalloids, which are toxic to most plant species. There are three main responses in plants induced by excessive elemental concentrations in the substrate: exclusion, indication, and accumulation. At the extreme end of accumulation are hyperaccumulator plant species, whose elemental concentrations in the leaves are 2–3 orders of magnitude higher than the concentrations in the leaves of nonaccumulators, mainly sequestered in the epidermal tissues of the leaves.

The studies of metallophytes provide important insights into the adaptation and evolution of plants under extreme environmental stress. Their existence on metalliferous soils also contributes to the understanding of natural selection and local adaptation as well as the broader impact of plant biodiversity in challenging habitats. In addition, these plants have significant potential for practical applications, particularly in phytoremediation, which has been increasingly used in agromining in recent decades. The ability of metallophytes to restore vegetation cover and ecosystem function in disturbed and contaminated areas also contributes to ecological restoration. By further exploring and utilizing these plants, we can better protect natural ecosystems, rehabilitate polluted landscapes, and promote a more sustainable interaction between human industry and the natural world.

References

Adamo, P., Denaix, L., Terribile, F., Zampella, M., 2003. Characterization of heavy metals in contaminated volcanic soils of the Solofrana river valley (southern Italy). Geoderma 117 (3–4), 347–366. Available at https://doi.org/10.1016/s0016–7061(03)00133-2.

Ahmad, P., Ahanger, M.A., Alyemeni, M.N., Alam, P., Dari, B., Sihi, D., et al., 2019. Heavy metals and photosynthesis: recent developments. Photosynthesis, Productivity and Environmental Stress. Wiley, pp. 107–134. Available at http://doi.org/10.1002/9781119501800.ch7.

Antonovics, J., Bradshaw, A.D., Turner, R.G., 1971. Heavy metal tolerance in plants. Advances in Ecological Research 7 (C), 1–85. Available at https://doi.org/10.1016/S0065-2504(08)60202-0.

Alford, E.R., Pilon-Smits, E.A.H., Paschke, M.W., 2010. Metallophytes: a view from the rhizosphere. Plant and Soil 337 (1), 33–50. Available at https://doi.org/10.1007/s11104-010-0482-3.

Assunção, A.G.L., Martins, P.D.C., De Folter, S., Vooijs, R., Schat, H., Aarts, M.G.M., 2001. Elevated expression of metal transporter genes in three accessions of the metal hyperaccumulator *Thlaspi caerulescens*. Plant, Cell & Environment 24 (2), 217–226. Available at https://doi.org/10.1111/j.1365-3040.2001.00666.x.

Bačeva Andonovska, K., Stafilov, T., Matevski, V., 2021. Accumulation abilities of endemic plant species from the vicinity of an As-Sb-Tl abandoned mine, Allchar, Kožuf Mountain. In: Balabanova, B., Stafilov, T. (Eds.), Contaminant Levels and Ecological Effects. Springer, Cham, pp. 375–402. Available at https://doi.org/10.1007/978-3-030-66135-9_13.

Badea, D.N., 2015. Determination of potentially toxic heavy metals (Pb, Hg, Cd) in popular medicinal herbs in the coal power plant area. Revista de Chimie 66 (8), 1132–1136. Available at http://www.revistadechimie.ro/archive.asp.

Baek, S.A., Han, T., Ahn, S.K., Kang, H., Cho, M.R., Lee, S.C., et al., 2012. Effects of heavy metals on plant growths and pigment contents in *Arabidopsis thaliana*. Plant Pathology Journal 28 (4), 446–452. Available at https://doi.org/10.5423/PPJ.NT.01.2012.0006.

Baker, A.J.M., 1981. Accumulators and excluders: strategies in the response of plants to heavy metals. Journal of Plant Nutrition 3 (1-4), 643–654. Available at https://doi.org/10.1080/01904168109362867.

Baker, A.J.M., 1987. Metal tolerance. New Phytologist 106, 93–111. Available at https://doi.org/10.1111/j.1469-8137.1987.tb04685.x.

Baker, A.J.M., Brooks, R., 1989. Terrestrial higher plants which hyperaccumulate metallic elements. A review of their distribution, ecology and phytochemistry. Biorecovery 1 (2), 81–126.

Baker, A.J.M., Ernst, W.H.O., van der Ent, A., Malaisse, F., Ginocchio, R., 2010. Metallophytes: The Unique Biological Resource, Its Ecology and Conservational Status in Europe, Central Africa and Latin America. Cambridge University Press (CUP), pp. 7–40. Available at http://doi.org/10.1017/cbo9780511805561.003.

Banásová, V., Horak, O., Čiamporová, M., Nadubinská, M., Lichtscheidl, I., 2006. The vegetation of metalliferous and non-metalliferous grasslands in two former mine regions in Central Slovakia. Biologia (Lahore, Pakistan) 61 (4), 433–439. Available at https://doi.org/10.2478/s11756-006-0073-1.

Baker, A.J.M., Proctor, J., van Balgooy, M.M.J., Reeves, R.D., 1992. Hyperaccumulation of nickel by the flora of the ultramafics of Palawan, Republic of the Philippines. In: Baker, A.J.M., Proctor, J., Reeves, R.D. (Eds.), The Vegetation of Ultramafic (Serpentine) Soils. Intercept Ltd, Andover (UK), pp. 291–304.

Baker, A.J.M., Whiting, S.N., 2002. In search of the Holy Grail: a further step in understanding metal hyperaccumulation? New Phytologist 155 (1), 1–4. Available at https://doi.org/10.1046/j.1469-8137.2002.00449_1.x.

Baker, A.J.M., Whiting, S.N., 2008. Metallophytes—a unique biodiversity and biotechnological resource in the care of the minerals industry. In: Fourie, A.B., Tibbett, M., Weiersbye, I., Dye, P. (Eds.), Mine Closure 2008: Proceedings of the Third International Seminar on Mine Closure. Australian Centre for Geomechanics, Perth, pp. 13–20. Available at https://doi.org/10.36487/ACG_repo/852_2.

Barceló, J., Vázquez, M.D., Poschenrieder, C., 1988. Structural and ultrastructural disorders in cadmium-treated bush bean plants (*Phaseolus vulgaris* L). New Phytologist 108 (1), 37–49. Available at https://doi.org/10.1111/j.1469-8137.1988.tb00202.x.

Baumbach, H., Volkmann, H.K.M., Wolkersdorfer, C., 2007. Schwermetallrasen auf Hüttenstäuben am Weinberg bei Hettstedt-Burgörner (Mansfelder Land) - Ergebnis jahrhundertelanger Kontamination und Herausforderung für den Naturschutz. Hercynia 40 (1), 87–109.

Behmer, S.T., Lloyd, C.M., Raubenheimer, D., Stewart-Clark, J., Knight, J., Leighton, R.S., et al., 2005. Metal hyperaccumulation in plants: mechanisms of defence against insect herbivores. Functional Ecology 19 (1), 55–66. Available at https://doi.org/10.1111/j.0269-8463.2005.00943.x.

Belzile, N., Chen, Y.W., 2015. Tellurium in the environment: a critical review focused on natural waters, soils, sediments and airborne particles. Applied Geochemistry 63, 83–92. Available at https://doi.org/10.1016/j.apgeochem.2015.07.002.

Blissett A.H., 1966. Copper tolerant plants from the Ukaparinga copper mine, Williamstown. *Quarterly Geological Notes, Geological Survey of South Australia* 18, 1–4.

Bothe, H., 2011. Plants in heavy metal soils. In: Sherameti, I., Varma, A. (Eds.), Detoxification of Heavy Metals. Soil Biology 30. Springer, Berlin, Heidelberg, pp. 35–57. Available at https://doi.org/10.1007/978-3-642-21408-0_2.

Boyd, R.S., 2007. The defense hypothesis of elemental hyperaccumulation: status, challenges and new directions. Plant and Soil 293 (1–2), 153–176. Available at https://doi.org/10.1007/s11104-007-9240-6.

Boyd, R.S., 2014. Ecology and evolution of metal-hyperaccumulating plants. In: Rajakaruna, N., Boyd, R.S., Harris, T.B. (Eds.), Plant Ecology and Evolution in Harsh Environment. Nova Science Publishers, Hauppauge, pp. 227–241.

Bradshaw, A.D., 1984. Adaptation of plants to soils containing toxic metals: a test for conceit. In: In: Evered, D., Collins, G.M. (Eds.), Origins and Development of Adaptation 102. CIBA Foundation Symposium, Pitman, London, pp. 4–19. Available at https://doi.org/10.1002/9780470720837.

Brooks, R.R., 1994. Plants that hyperaccumulate heavy metals. In: Farago, M. (Ed.), Plants and the Chemical Elements: Biochemistry, Uptake, Tolerance and Toxicity. VCH, Weinheim, pp. 87–105. Available at https://doi.org/10.1002/9783527615919.

Bărbulescu, A., Barbeş, L., Dumitriu, C.Ş., 2022. Impact of soil pollution on melliferous plants. Toxics 10 (5), 239. Available at https://doi.org/10.3390/toxics10050239.

Cambrollé, J., García, J.L., Figueroa, M.E., Cantos, M., 2015. Evaluating wild grapevine tolerance to copper toxicity. Chemosphere 120, 171–178. Available at https://doi.org/10.1016/j.chemosphere.2014.06.044.

Cappa, J.J., Pilon-Smits, E.A.H., 2014. Evolutionary aspects of elemental hyperaccumulation. Planta 239 (2), 267–275. Available at https://doi.org/10.1007/s00425-013-1983-0.

Chiarucci, A., Maccherini, S., Bonini, I., De Dominicis, V., 1999. Effects of nutrient addition on community productivity and structure of serpentine vegetation. Plant Biology 1 (01), 121–126. Available at https://doi.org/10.1055/s-2007-978498.

Choudhary, M., Jetley, U.K., Abash Khan, M., Zutshi, S., Fatma, T., 2007. Effect of heavy metal stress on proline, malondialdehyde, and superoxide dismutase activity in the cyanobacterium *Spirulina platensis*-S5. Ecotoxicology and Environmental Safety 66 (2), 204–209. Available at https://doi.org/10.1016/j.ecoenv.2006.02.002.

Corzo Remigio, A., Pošćić, F., Nkrumah, P.N., Edraki, M., Spiers, K.M., Brueckner, D., et al., 2022. Comprehensive insights in thallium ecophysiology in the hyperaccumulator *Biscutella laevigata*. Science of the Total Environment 838 (Pt 2), 155899. Available at https://doi.org/10.1016/j.scitotenv.2022.155899.

Đorđević, T., Drahota, P., Kolitsch, U., Majzlan, J., Peřestá, M., Kiefer, S., et al., 2021. Synergetic Tl and As retention in secondary minerals: an example of extreme arsenic and thallium pollution. Applied Geochemistry 135, 105114. Available at https://doi.org/10.1016/j.apgeochem.2021.105114.

Deng, T.H.B., Tang, Y.T., van der Ent, A., Sterckeman, T., Echevarria, G., Morel, J.L., et al., 2016. Nickel translocation via the phloem in the hyperaccumulator *Noccaea caerulescens* (Brassicaceae). Plant and Soil 404 (1–2), 35–45. Available at https://doi.org/10.1007/s11104-016-2825-1.

Deng, T.H.B., van der Ent, A., Tang, Y.T., Sterckeman, T., Echevarria, G., Morel, J.L., et al., 2018. Nickel hyperaccumulation mechanisms: a review on the current state of knowledge. Plant and Soil 423 (1–2), 1–11. Available at https://doi.org/10.1007/s11104-017-3539-8.

Dinh, N.T., Vu, D.T., Mulligan, D., Nguyen, A.V., 2015. Accumulation and distribution of zinc in the leaves and roots of the hyperaccumulator *Noccaea caerulescens*. Environmental and Experimental Botany 110, 85–95. Available at https://doi.org/10.1016/j.envexpbot.2014.10.001.

El Desouky, H.A., Muchez, P., Cailteux, J., 2009. Two Cu-Co sulfide phases and contrasting fluid systems in the Katanga Copperbelt, Democratic Republic of Congo. Ore Geology Reviews 36 (4), 315–332. Available at https://doi.org/10.1016/j.oregeorev.2009.07.003.

El Mehdawi, A.F., Quinn, C.F., Pilon-Smits, E.A.H., 2011a. Selenium hyperaccumulators facilitate selenium-tolerant neighbors via phytoenrichment and reduced herbivory. Current Biology 21 (17), 1440–1449. Available at https://doi.org/10.1016/j.cub.2011.07.033.

El Mehdawi, A.F., Quinn, C.F., Pilon-Smits, E.A.H., 2011b. Effects of selenium hyperaccumulation on plant-plant interactions: Evidence for elemental allelopathy? New Phytologist 191 (1), 120–131. Available at https://doi.org/10.1111/j.1469-8137.2011.03670.x.

Ernst, W., 1976. Physiological and biochemical aspects of metal tolerance. In: Mansfield, T.A. (Ed.), Effects of Air Pollutants on Plants. Cambridge University Press, Cambridge, pp. 115–133.

Ernst, W.H.O., 1990. Mine vegetation in Europe. Heavy Metal Tolerance in Plants: Evolutionary Aspects 18. pp. 21–38.

Ernst, W.H.O., 2000. Evolution of metal hyperaccumulation and phytoremediation hype. New Phytologist 146 (3), 357–358. Available at https://doi.org/10.1046/j.1469-8137.2000.00669.x.

Faucon, M.P., Shutcha, M.N., Meerts, P., 2007. Revisiting copper and cobalt concentrations in supposed hyperaccumulators from SC Africa: influence of washing and metal concentrations in soil. Plant and Soil 301 (1–2), 29–36. Available at https://doi.org/10.1007/s11104-007-9405-3.

Favas, P.J., Pratas, J., Varun, M., D'Souza, R., Paul, M.S., 2014. Phytoremediation of soils contaminated with metals and metalloids at mining areas: potential of native flora. In: In: Hernandez-Soriano (Ed.), Environmental Risk Assessment of Soil Contamination 3. IntechOpen, pp. 485–516.

Favero-Longo, S.E., 2014. Lichens on metal-rich substrates. In: Rajakaruna, N., Boyd, R.S., Harris, T.B. (Eds.), Plant Ecology and Evolution in Harsh Environments. Nova Publishers, pp. 53–76.

Favero-Longo, S.E., Isocrono, D., Piervittori, R., 2004. Lichens and ultramafic rocks: a review. The Lichenologist 36 (6), 391–404. Available at https://doi.org/10.1017/s0024282904014215.

Fernando, D.R., Woodrow, I.E., Jaffré, T., Dumontet, V., Marshall, A.T., Baker, A.J.M., 2008. Foliar manganese accumulation by *Maytenus founieri* (Celastraceae) in its native New Caledonian habitats: populational variation and localization by X-ray microanalysis. New Phytologist 177 (1), 178–185. Available at https://doi.org/10.1111/j.1469-8137.2007.02253.x.

Frérot, H., Lefèbvre, C., Gruber, W., Collin, C., Santos, A.D., Escarré, J., 2006. Specific interactions between local metallicolous plants improve the phytostabilization of mine soils. Plant and Soil 282 (1–2), 53–65. Available at https://doi.org/10.1007/s11104-005-5315-4.

Galeas, M.L., Zhang, L.H., Freeman, J.L., Wegner, M., Pilon-Smits, E.A.H., 2007. Seasonal fluctuations of selenium and sulfur accumulation in selenium hyperaccumulators and related nonaccumulators. New Phytologist 173 (3), 517–525. Available at https://doi.org/10.1111/j.1469-8137.2006.01943.x.

Galey, M.L., van der Ent, A., Iqbal, M.C.M., Rajakaruna, N., 2017. Ultramafic geoecology of South and Southeast Asia. Botanical Studies 58 (1), 18. Available at https://doi.org/10.1186/s40529-017-0167-9.

Gei, V., Isnard, S., Erskine, P.D., Echevarria, G., Fogliani, B., Jaffré, T., et al., 2020. A systematic assessment of the occurrence of trace element hyperaccumulation in the flora of New Caledonia. Botanical Journal of the Linnean Society 194 (1), 1–22. Available at https://doi.org/10.1093/botlinnean/boaa029.

Goudard, L., Blaudez, D., Sirguey, C., Purwadi, I., Invernon, V., Rouhan, G., et al., 2024. Prospecting for rare earth element (hyper)accumulators in the Paris Herbarium using X-ray fluorescence spectroscopy reveals new distributional and taxon discoveries. Annals of Botany 133 (4), 573–584. Available at https://doi.org/10.1093/aob/mcae011.

Harvey, M.A., Erskine, P.D., Harris, H.H., Brown, G.K., Pilon-Smits, E.A.H., Casey, L.W., et al., 2020. Distribution and chemical form of selenium in *Neptunia amplexicaulis* from Central Queensland, Australia. Metallomics 12 (4), 514–527. Available at https://doi.org/10.1039/c9mt00244h.

Holyoak, D.T., 2008. Bryophytes and metallophyte vegetation on metalliferous mine-waste in Ireland: report to National Parks and Wildlife Service of a survey in 2008. Unpublished Report of the National Parks and Wildlife Service.

Holyoak, D.T., Lockhart, N., 2011. A survey of bryophytes and metallophyte vegetation of metalliferous mine spoil in Ireland. Journal of the Mining Heritage Trust of Ireland 11, 3–16.

Huneck, S., 2006. Die Flechten der Kupferschieferhalden um Eisleben, Mansfeld und Sangerhausen. Mitteilungen zur floristischen Kartierung in Sachsen-Anhalt, *Sonderheft* 4 (11), 1–62.

Jaffré, T., Brooks, R.R., Lee, J., Reeves, R.D., 1976. *Sebertia acuminata*: a hyperaccumulator of nickel from New Caledonia. Science (New York, N.Y.) 193 (4253), 579–580. Available at https://doi.org/10.1126/science.193.4253.579.

Jaffré, T., Pillon, Y., Thomine, S., Merlot, S., 2013. The metal hyperaccumulators from New Caledonia can broaden our understanding of nickel accumulation in plants. Frontiers in Plant Science 4, 279. Available at https://doi.org/10.3389/fpls.2013.00279.

Jakovljević, K., Bani, A., Pavlova, D., Konstantinou, M., Dimitrakopoulos, P.G., Kyrkas, D., et al., 2022. Hyperaccumulator plant discoveries in the Balkans: accumulation, distribution, and practical applications. Botanica Serbica 46 (2), 161–178. Available at https://doi.org/10.2298/BOTSERB2202161J, http://botanicaserbica.bio.bg.ac.rs/.

Jakovljević, K., Lakušić, D., Vukojičić, S., Tomović, G., Šinžar-Sekulić, J., Stevanović, V., 2011. Richness and diversity of Pontic flora on serpentine of Serbia. Central European Journal of Biology 6 (2), 260–274. Available at https://doi.org/10.2478/s11535-010-0110-5.

Jakovljević, K., Bačeva Andonovska, K., Salinitro, M., Mišljenović, T., van der Ent, A., 2025. Biogeochemical survey of the Allchar (North Macedonia) arsenic-thallium ore body: a focus on hyperaccumulator plants. Plant and Soil 513 (1). Available at https://doi.org/10.1007/s11104-025-07252-6.

Jakovljević, K., Mišljenović, T., Bačeva Andonovska, K., Echevarria, G., Baker, A.J.M., Brueckner, D., et al., 2023. Thallium hyperaccumulation status of the violets of the Allchar arsenic-thallium deposit (North Macedonia) confirmed through synchrotron μXRF imaging. Metallomics 15 (11). Available at https://doi.org/10.1093/mtomcs/mfad063.

Jakovljević, K., Mišljenović, T., Bačeva Andonovska, K., Echevarria, G., Charrois, L., van der Ent, A., 2024. Living at the edge of life: metallophytes from the most toxic arsenic-thallium tailings in the world (Allchar, North Macedonia). Plant and Soil 497 (1–2), 413–428. Available at https://doi.org/10.1007/s11104-023-06404-w.

Jakovljević, K., Mišljenović, T., Savović, J., Ranković, D., Ranđelović, D., Mihailović, N., et al., 2020. Accumulation of trace elements in *Tussilago farfara* colonizing post-flotation tailing sites in Serbia. Environmental Science and Pollution Research 27 (4), 4089–4103. Available at https://doi.org/10.1007/s11356-019-07010-z.

Jakovljević, K., Mišljenović, T., van der Ent, A., Baker, A.J.M., Andrejić, G., Tomović, G., et al., 2023. Zinc (hyper)accumulation in *Cardamine waldsteinii* from discovery in the herbarium to validation in the field. Plant Biosystems - An International Journal Dealing with all Aspects of Plant Biology 157 (4), 851–857. Available at https://doi.org/10.1080/11263504.2023.2204318.

Jakovljević, K., Mišljenović, T., van der Ent, A., Baker, A.J.M., Invernón, V.R., Echevarria, G., 2024. Mining" the herbarium for hyperaccumulators: discoveries of nickel and zinc (hyper)accumulation in the genus *Noccaea* (Brassicaceae) through X-ray fluorescence herbarium scanning. Ecological Research 39 (4), 450–459. Available at https://doi.org/10.1111/1440-1703.12448.

Jenny, H., 1980. The soil resource: origin and behavior. Ecological Studies 37, 1–376.

Jhee, E.M., Boyd, R.S., Eubanks, M.D., Davis, M.A., 2006. Nickel hyperaccumulation by *Streptanthus polygaloides* protects against the folivore *Plutella xylostella* (Lepidoptera: Plutellidae). Plant Ecology 183 (1), 91–104. Available at https://doi.org/10.1007/s11258-005-9009-z.

Jin, X., Yang, X., Islam, E., Liu, D., Mahmood, Q., 2008. Effects of cadmium on ultrastructure and antioxidative defense system in hyperaccumulator and non-hyperaccumulator ecotypes of *Sedum alfredii* Hance. Journal of Hazardous Materials 156 (1-3), 387–397. Available at https://doi.org/10.1016/j.jhazmat.2007.12.064.

Johnston, W.R., Proctor, J., 1977. A comparative study of metal levels in plants from two contrasting lead-mine sites. Plant and Soil 46 (1), 251–257. Available at https://doi.org/10.1007/BF00693131.

Kazakou, E., Adamidis, G.C., Baker, A.J.M., Reeves, R.D., Godino, M., Dimitrakopoulos, P.G., 2010. Species adaptation in serpentine soils in Lesbos Island (Greece): metal hyperaccumulation and tolerance. Plant and Soil 332 (1), 369–385. Available at https://doi.org/10.1007/s11104-010-0302-9.

Kazakou, E., Dimitrakopoulos, P.G., Baker, A.J.M., Reeves, R.D., Troumbis, A.Y., 2008. Hypotheses, mechanisms and trade-offs of tolerance and adaptation to serpentine soils: from species to ecosystem level. Biological Reviews 83 (4), 495–508. Available at https://doi.org/10.1111/j.1469-185X.2008.00051.x.

Khan, A.M., Bakar, N.K.A., Bakar, A.F.A., Ashraf, M.A., 2017. Chemical speciation and bioavailability of rare earth elements (REEs) in the ecosystem: a review. Environmental Science and Pollution Research 24 (29), 22764–22789. Available at https://doi.org/10.1007/s11356-016-7427-1.

Khanna, K., Jamwal, V.L., Gandhi, S.G., Ohri, P., Bhardwaj, R., 2019. Metal resistant PGPR lowered Cd uptake and expression of metal transporter genes with improved growth and photosynthetic pigments in *Lycopersicon esculentum* under metal toxicity. Scientific Reports 9 (1). Available at https://doi.org/10.1038/s41598-019-41899-3, www.nature.com/srep/index.html.

Kierczak, J., Pietranik, A., Pędziwiatr, A., 2021. Ultramafic geoecosystems as a natural source of Ni, Cr, and Co to the environment: a review. Science of the Total Environment 755 (Pt 1), 142620. Available at https://doi.org/10.1016/j.scitotenv.2020.142620.

Kozhevnikova, A.D., Seregin, I.V., Verweij, R., Schat, H., 2014. Histidine promotes the loading of nickel and zinc, but not of cadmium, into the xylem in *Noccaea caerulescens*. Plant Signaling & Behavior 9 (9), e29580. Available at https://doi.org/10.4161/psb.29580.

Kruckeberg, A.R., 1954. The ecology of serpentine soils: III. Plant species in relation to serpentine soils. Ecology 35, 267–274.

Kruckeberg, A.R., 1992. Plant life of western North American ultramafics. In: In: Roberts, B.A., Proctor, J. (Eds.), The Ecology of Areas with Serpentinized Rocks. Geobotany 17. Springer, Dordrecht, pp. 31–73. Available at https://doi.org/10.1007/978-94-011-3722-5_3.

Kruckeberg, A.R., 2002. Geology and Plant Life: The Effects of Landforms and Rock Types on Plants. University of Washington Press, Seattle.

Krämer, U., 2010. Metal hyperaccumulation in plants. Annual Review of Plant Biology 61 (1), 517–534. Available at https://doi.org/10.1146/annurev-arplant-042809-112156.

Krämer, U., Cotter-Howells, J.D., Charnock, J.M., Baker, A.J.M., Smith, J.A.C., 1996. Free histidine as a metal chelator in plants that accumulate nickel. Nature 379 (6566), 635–638. Available at https://doi.org/10.1038/379635a0.

Küpper, H., Lombi, E., Zhao, F.J., Wieshammer, G., McGrath, S.P., 2001. Cellular compartmentation of nickel in the hyperaccumulators Alyssum lesbiacum, Alyssum bertolonii and Thlaspi goesingense. Journal of Experimental Botany 52 (365), 2291–2300. Available at https://doi.org/10.1093/jexbot/52.365.2291.

LaCoste, C., Robinson, B., Brooks, R., Anderson, C., Chiarucci, A., Leblanc, M., 1999. The phytoremediation potential of thallium-contaminated soils using Iberis and *Biscutella* species. International Journal of Phytoremediation 1 (4), 327–338. Available at https://doi.org/10.1080/15226519908500023.

Leyval, C., Turnau, K., Haselwandter, K., 1997. Effect of heavy metal pollution on mycorrhizal colonization and function: physiological, ecological and applied aspects. Mycorrhiza 7 (3), 139–153. Available at https://doi.org/10.1007/s005720050174.

Li, Y., Zhang, S., Jiang, W., Liu, D., 2013. Cadmium accumulation, activities of antioxidant enzymes, and malondialdehyde (MDA) content in *Pistia stratiotes* L. Environmental Science and Pollution Research 20 (2), 1117–1123. Available at https://doi.org/10.1007/s11356-012-1054-2.

Liang, X., Zhang, L., Natarajan, S.K., Becker, D.F., 2013. Proline mechanisms of stress survival. Antioxidants and Redox Signaling 19 (9), 998–1011. Available at https://doi.org/10.1089/ars.2012.5074.

Likar, M., Pongrac, P., Vogel-Mikuš, K., Regvar, M., 2010. Molecular diversity and metal accumulation of different *Thlaspi praecox* populations from Slovenia. Plant and Soil 330 (1–2), 195–205. Available at https://doi.org/10.1007/s11104-009-0192-x.

Liu, C., Liu, W.S., van der Ent, A., Morel, J.L., Zheng, H.X., Wang, G.B., et al., 2021. Simultaneous hyperaccumulation of rare earth elements, manganese and aluminum in *Phytolacca americana* in response to soil properties. Chemosphere 282, 131096. Available at https://doi.org/10.1016/j.chemosphere.2021.131096.

Liu, J., Luo, X., Sun, Y., Tsang, D.C.W., Qi, J., Zhang, W., et al., 2019. Thallium pollution in China and removal technologies for waters: a review. Environment International 126, 771–790. Available at https://doi.org/10.1016/j.envint.2019.01.076.

Lombi, E., Zhao, F.J., McGrath, S.P., Young, S.D., Sacchi, G.A., 2001. Physiological evidence for a high-affinity cadmium transporter highly expressed in a *Thlaspi caerulescens* ecotype. New Phytologist 149 (1), 53–60. Available at https://doi.org/10.1046/j.1469-8137.2001.00003.x.

Manca, P.P., Massacci, G., Pintus, D., Sogos, G., 2021. The flotation of sphalerite mine tailings as a remediation method. Minerals Engineering 165, 106862. Available at https://doi.org/10.1016/j.mineng.2021.106862.

Mari, S., Gendre, D., Pianelli, K., Ouerdane, L., Lobinski, R., Briat, J.-F., et al., 2006. Root-to-shoot long-distance circulation of nicotianamine and nicotianamine-nickel chelates in the metal hyperaccumulator *Thlaspi caerulescens*. Journal of Experimental Botany 57 (15), 4111–4122. Available at https://doi.org/10.1093/jxb/erl184.

McNear, D.H., Peltier, E., Everhart, J., Chaney, R.L., Sutton, S., Newville, M., et al., 2005. Application of quantitative fluorescence and absorption-edge computed microtomography to image metal compartmentalization in *Alyssum murale*. Environmental Science and Technology 39 (7), 2210–2218. Available at https://doi.org/10.1021/es0492034.

Mickovski, S.B., van Beek, L.P.H., 2009. Root morphology and effects on soil reinforcement and slope stability of young vetiver (*Vetiveria zizanioides*) plants grown in semi-arid climate. Plant and Soil 324 (1), 43–56. Available at https://doi.org/10.1007/s11104-009-0130-y.

Nagy, L., Proctor, J., 1997. Plant growth and reproduction on a toxic alpine ultramafic soil: adaptation to nutrient limitation. New Phytologist 137 (2), 267–274. Available at https://doi.org/10.1046/j.1469-8137.1997.00799.x.

Naing, A.H., Kim, C.K., 2021. Abiotic stress-induced anthocyanins in plants: their role in tolerance to abiotic stresses. Physiologia Plantarum 172 (3), 1711–1723. Available at https://doi.org/10.1111/ppl.13373.

O'Dell, R.E., Claassen, V.P., 2006. Serpentine and nonserpentine *Achillea millefolium* accessions differ in serpentine substrate tolerance and response to organic and inorganic amendments. Plant and Soil 279 (1–2), 253–269. Available at https://doi.org/10.1007/s11104-005-2360-y.

O'Dell, R.E., James, J.J., Richards, J.H., 2006. Congeneric serpentine and nonserpentine shrubs differ more in leaf Ca:Mg than in tolerance of low N, low P, or heavy metals. Plant and Soil 280 (1–2), 49–64. Available at https://doi.org/10.1007/s11104-005-3502-y.

Paçarizi, M., Krasniqi, E., Zogaj, M., & Ukaj, S., 2020. Hyper-accumulation of nickel by *Bornmuellera dieckii* Degen as an endemic plant species in Kosovo. In: Stanojević, J., Kerolli-Mustafa, M. (eds), International Scientific Conference on Theory to Practice as a Cognitive, Educational and Social Challenge, 17–18 September 2020, Mitrovica, Kosovo, pp. 277−284.

Pereira, W.E., De Siqueira, D.L., MartÍnez, C.A., Puiatti, M., 2000. Gas exchange and chlorophyll fluorescence in four citrus rootstocks under aluminium stress. Journal of Plant Physiology 157 (5), 513–520. Available at https://doi.org/10.1016/S0176-1617(00)80106-6.

Piervittori, R., Isocrono, D., Favero Longo, S.E., De Nicolo, A., 2004. Indagini floristiche ed ecologiche sui licheni degli" ambienti ofiolitici" del Parco Naturale del Mont Avic: influenza della natura geologica del substrato sulle comunità licheniche rupicole e terricole. Revue Valdotaine D'Histoire Naturelle 58, 51–64.

Pollard, A.J., Reeves, R.D., Baker, A.J.M., 2014. Facultative hyperaccumulation of heavy metals and metalloids. Plant Science 217–218, 8–17. Available at https://doi.org/10.1016/j.plantsci.2013.11.011.

Ranđelović, D., 2023. Aromatic herbs and shrubs in phytoremediation. In: Pandey, V.C., Ancona, V., Roy, M., Ranđelović, D. (Eds.), Aromatic Plant-Based Phytoremediation, Socio-Economic and Agricultural Sustainability. Elsevier, pp. 75–124.

Ranđelović, D., Jakovljević, K., Mihailović, N., Jovanović, S., 2018. Metal accumulation in populations of *Calamagrostis epigejos*(L) Roth from diverse anthropogenically degraded sites (SE Europe, Serbia). Environmental Monitoring and Assessment 190 (4), 183. Available at https://doi.org/10.1007/s10661-018-6514-9.

Rascio, N., Navari-Izzo, F., 2011. Heavy metal hyperaccumulating plants: how and why do they do it? And what makes them so interesting? Plant Science 180 (2), 169–181. Available at https://doi.org/10.1016/j.plantsci.2010.08.016.

Raskin, I., Kumar, P.N., Dushenkov, S., Salt, D.E., 1994. Bioconcentration of heavy metals by plants. Current Opinion in Biotechnology 5 (3), 285–290. Available at https://doi.org/10.1016/0958-1669(94)90030-2.

Reeves, R.D., 2003. Tropical hyperaccumulators of metals and their potential for phytoextraction. Plant and Soil 249 (1), 57–65. Available at https://doi.org/10.1023/a:1022572517197.

Reeves, R.D., Baker, A.J.M., 1984. Studies on metal uptake by plants from serpentine and non-serpentine populations of *Thlaspi goesingense* Hálácsy (Cruciferae). New Phytologist 98 (1), 191–204. Available at https://doi.org/10.1111/j.1469-8137.1984.tb06108.x.

Reeves, R.D., Baker, A.J.M., 2000. Metal accumulating plants. In: Raskin, I., Ensley, B. (Eds.), Phytoremediation of Toxic Metals: Using Plants to Clean Up the Environment. Wiley, New York, pp. 193–229.

Reeves, R.D., Baker, A.J.M., Borhidi, A., Berazaín, R., 1996. Nickel-accumulating plants from the ancient serpentine soils of Cuba. New Phytologist 133 (2), 217–224. Available at https://doi.org/10.1111/j.1469-8137.1996.tb01888.x.

Reeves, R.D., Baker, A.J.M., Jaffré, T., Erskine, P.D., Echevarria, G., van der Ent, A., 2018. A global database for plants that hyperaccumulate metal and metalloid trace elements. New Phytologist 218 (2), 407–411. Available at https://doi.org/10.1111/nph.14907.

Reeves, R.D., Baker, A.J.M., Kelepertsis, A., 1997. The distribution and biogeochemistry of some serpentine plants of Greece. In: Jaffré, T., Reeves, R.D., Becquer, T. (Eds.), Écologie des milieux sur roches ultramafi ques et sur sols métallifères ORSTOM, Nouméa, pp. 205–207 Documents Scientifiques et Techniques No. III/2.

Reeves, R.D., Brooks, R.R., 1983. European species of *Thlaspi*L. (Cruciferae) as indicators of nickel and zinc. Journal of Geochemical Exploration 18 (3), 275–283. Available at https://doi.org/10.1016/0375-6742(83)90073-0.

Reeves, R.D., Brooks, R.R., Dudley, T.R., 1983. Uptake of nickel by species of *Alyssum*, *Bornmuellera*, and other genera of Old World Tribus Alysseae. Taxon 32 (2), 184–192. Available at https://doi.org/10.2307/1221970.

Reeves, R.D., Schwartz, C., Morel, J.L., Edmondson, J., 2001. Distribution and metal-accumulating behavior of *Thlaspi caerulescens* and associated metallophytes in France. International Journal of Phytoremediation 3 (2), 145–172. Available at https://doi.org/10.1080/15226510108500054.

Reeves, R.D., van der Ent, A., Echevarria, G., Isnard, S., Baker, A.J.M., 2021. Global distribution and ecology of hyperaccumulator plants. In: van der Ent, A., Baker, A.J.M., Echevarria, G., Simonnot, M.O., Morel, J.L. (Eds.), Agromining: Farming for Metals. Mineral Resource Reviews. Springer, Cham, pp. 133–154. Available at https://doi.org/10.1007/978-3-030-58904-2_7.

Robinson, B.H., Lombi, E., Zhao, F.J., McGrath, S.P., 2003. Uptake and distribution of nickel and other metals in the hyperaccumulator *Berkheya coddii*. New Phytologist 158 (2), 279–285. Available at https://doi.org/10.1046/j.1469-8137.2003.00743.x.

Robinson, B., Kim, N., Marchetti, M., Moni, C., Schroeter, L., van den Dijssel, C., et al., 2006. Arsenic hyperaccumulation by aquatic macrophytes in the Taupo Volcanic Zone, New Zealand. Environmental and Experimental Botany 58 (1-3), 206–215. Available at https://doi.org/10.1016/j.envexpbot.2005.08.004.

Rodwell, J.S., Morgan, V., Jefferies, R.G., Moss, D., 2007. The European context of the British lowland grassland. JNCC No. 349, Chapter 7. Metallophyte Vegetation. Joint Nature Conservation Committee, Peterborough, UK.

Rue, M., Paul, A.L.D., Echevarria, G., van der Ent, A., Simonnot, M.O., Morel, J.L., 2020. Uptake, translocation and accumulation of nickel and cobalt in *Berkheya coddii*, a 'metal crop' from South Africa. Metallomics 12 (8), 1278–1289. Available at https://doi.org/10.1039/d0mt00099j.

Schat, H., Sharma, S.S., Vooijs, R., 1997. Heavy metal-induced accumulation of free proline in a metal-tolerant and a nontolerant ecotype of *Silene vulgaris*. Physiologia Plantarum 101 (3), 477–482. Available at https://doi.org/10.1111/j.1399-3054.1997.tb01026.x.

Seregin, I.V., Kozhevnikova, A.D., Schat, H., 2021. Correlated variation of the Zn accumulation and tolerance capacities among populations and ecotypes of the Zn hyperaccumulator, *Noccaea caerulescens*. Russian Journal of Plant Physiology 68, S26–S36. Available at https://doi.org/10.1134/S1021443721070128.

Sharma, S., Joshi, J., Kataria, S., Verma, S.K., Chatterjee, S., Jain, M., et al., 2020. Regulation of the Calvin cycle under abiotic stresses: an overview. In: Tripathi, D.K., Singh, V.P., Chauhan, D.K., Sharma, S., Prasad, S.M., Dubey, N.K., Ramawat, N. (Eds.), Plant Life Under Changing Environment. Academic Press, pp. 681–717. Available at https://doi.org/10.1016/B978-0-12-818204-8.00030-8.

Souri, Z., Karimi, N., Sandalio, L.M., 2017. Arsenic hyperaccumulation strategies: an overview. Frontiers in Cell and Developmental Biology 5, 258555. Available at https://doi.org/10.3389/fcell.2017.00067.

Stefanowicz, A.M., Kapusta, P., Błońska, A., Kompała-Bąba, A., Woźniak, G., 2015. Effects of *Calamagrostis epigejos*, *Chamaenerion palustre* and *Tussilago farfara*on nutrient availability and microbial activity in the surface layer of spoil heaps after hard coal mining. Ecological Engineering 83, 328–337. Available at https://doi.org/10.1016/j.ecoleng.2015.06.034.

Stein, R.J., Höreth, S., de Melo, J.R.F., Syllwasschy, L., Lee, G., Garbin, M.L., et al., 2017. Relationships between soil and leaf mineral composition are element-specific, environment-dependent and geographically structured in the emerging model *Arabidopsis halleri*. New Phytologist 213 (3), 1274–1286. Available at https://doi.org/10.1111/nph.14219.

Stevanovic, V., Tan, K., Iatrou, G., 2003. Distribution of the endemic Balkan flora on serpentine I -obligate serpentine endemics. Plant Systematics and Evolution 242 (1–4), 149–170. Available at https://doi.org/10.1007/s00606-003-0044-8.

Sura-de Jong, M., Reynolds, R.J.B., Richterova, K., Musilova, L., Staicu, L.C., Chocholata, I., et al., 2015. Selenium hyperaccumulators harbor a diverse endophytic bacterial community characterized by high selenium resistance and plant growth promoting properties. Frontiers in Plant Science 6, 128592. Available at https://doi.org/10.3389/fpls.2015.00113.

Tang, Q., Tang, X., Li, Z., Chen, Y., Kou, N., Sun, Z., 2009. Adsorption and desorption behaviour of Pb(II) on a natural kaolin: equilibrium, kinetic and thermodynamic studies. Journal of Chemical Technology & Biotechnology 84 (9), 1371–1380. Available at https://doi.org/10.1002/jctb.2192.

Thiers, B., 2024. [continuously updated]. Index Herbariorum: a global directory of public herbaria and associated staff. New York Botanical Garden's Virtual Herbariumhttp://sweetgum.nybg.org/ih/

Vaculík, M., Mišljenović, T., Lukačová, Z., Jakovljević, K., Podar, D., Kováč, J., 2025. Zinc hyperaccumulation in plants: mechanisms and principles. In: Tripathi, D.K., Singh, V.P., Pandey, S., Sharma, S., Chauhan, D.K. (Eds.), Zinc in Plants. Current Knowledge and Recent Advances. Elsevier Academic Press., pp. 1–39. Available at https://doi.org/10.1016/B978-0-323-91314-0.00017-X.

van der Ent, A., Baker, A.J.M., Reeves, R.D., Chaney, R.L., Anderson, C.W.N., Meech, J.A., et al., 2015. Agromining: farming for metals in the future? Environmental Science and Technology 49 (8), 4773–4780. Available at https://doi.org/10.1021/es506031u, http://pubs.acs.org/journal/esthag.

van der Ent, A., Baker, A.J.M., Reeves, R.D., Pollard, A.J., Schat, H., 2013. Hyperaccumulators of metal and metalloid trace elements: facts and fiction. Plant and Soil 362 (1–2), 319–334. Available at https://doi.org/10.1007/s11104-012-1287-3.

van der Ent, A., Erskine, P., Sumail, S., 2015. Ecology of nickel hyperaccumulator plants from ultramafic soils in Sabah (Malaysia). Chemoecology 25 (5), 243–259. Available at https://doi.org/10.1007/s00049-015-0192-7.

van der Ent, A., Malaisse, F., Erskine, P.D., Mesjasz-Przybyłowicz, J., Przybyłowicz, W.J., Barnabas, A.D., et al., 2019. Abnormal concentrations of Cu-Co in: *Haumaniastrum katangense*, *Haumaniastrum robertii* and *Aeolanthus biformifolius*: contamination or hyperaccumulation? Metallomics 11 (3), 586–596. Available at https://doi.org/10.1039/c8mt00300a.

van der Ent, A., Pollard, J.A., Echevarria, G., Abubakari, F., Erskine, P.D., Baker, A.J.M., et al., 2021. Exceptional uptake and accumulation of chemical elements in plants: extending the hyperaccumulation paradigm. In: van der Ent, A., Baker, A.J.M., Echevarria, G., Simonnot, M.O., Morel, J.L. (Eds.), Agromining: Farming for Metals: Extracting Unconventional Resources Using Plants. Springer, Cham, pp. 99–131. Available at https://doi.org/10.1007/978-3-030-58904-2_6.

Visoottiviseth, P., Francesconi, K., Sridokchan, W., 2002. The potential of Thai indigenous plant species for the phytoremediation of arsenic contaminated land. Environmental Pollution 118 (3), 453–461. Available at https://doi.org/10.1016/s0269-7491(01)00293-7.

Wei, X., Zhou, Y., Tsang, D.C.W., Song, L., Zhang, C., Yin, M., et al., 2020. Hyperaccumulation and transport mechanism of thallium and arsenic in brake ferns (*Pteris vittata* L.): a case study from mining area. Journal of Hazardous Materials 388, 121756. Available at https://doi.org/10.1016/j.jhazmat.2019.121756.

Whittaker, R.H., 1954. The ecology of serpentine soils: IV. The vegetational response to serpentine soils. Ecology 35, 275–287. Available at https://www.jstor.org/stable/1931126.

Willems, J.H., 2004. Hoe is het eigenlijk met onze zink-vegetatie gesteld? Natuurhistorisch Maandblad 93, 21–25.

Wirth, V., 1972. Die Silikatflechten-Gemeinschaften in Ausseralpinen Zentraleuropa. Dissertationes Botanicae. Lehre 17, 1–305.

Wójcik, M., Gonnelli, C., Selvi, F., Dresler, S., Rostański, A., Vangronsveld, J., 2017. Metallophytes of serpentine and calamine soils: their unique ecophysiology and potential for phytoremediation. Advances in Botanical Research 83, 1–42. Available at https://doi.org/10.1016/bs.abr.2016.12.002.

Yang, W., Luo, L., Bostick, B.C., Wiita, E., Cheng, Y., Shen, Y., 2021. Effect of combined arsenic and lead exposure on their uptake and translocation in Indian mustard. Environmental Pollution 274, 116549. Available at https://doi.org/10.1016/j.envpol.2021.116549.

Zhao, F.J., Hamon, R.E., Lombi, E., McLaughlin, M.J., McGrath, S.P., 2002. Characteristics of cadmium uptake in two contrasting ecotypes of the hyperaccumulator *Thlaspi caerulescens*. Journal of Experimental Botany 53 (368), 535–543. Available at https://doi.org/10.1093/jexbot/53.368.535.

CHAPTER TWO

Metal-rich soils

2.1 Introduction

Soils are composed from variety of chemical elements, and those commonly present in small amounts are called trace elements. They are natural parts of soil and may include metals such as copper, cobalt, iron, manganese, molybdenum, and zinc. Some of them represent essential nutrients for living organisms in lower concentrations. However, when their content in soil is increased, they can be potentially harmful to biota. Metal-rich soils are those that contain elevated concentrations of metals or metalloids that may occur naturally (geogenic origin) or as a result of human activities (anthropogenic origin). Such soils are sometimes called metalifferous (meaning metal-containing). Increasing level of metal pollution worldwide contributes to the enlargement of the areas assigned to metalliferous soils. Distribution of metal-rich soils is confined to certain areas in the world known for existence of natural mineralization near or at the Earth's surface, as well as to the areas characterized by increased development of industry, urbanism, and agriculture. Processes like erosion or vertical distribution along the soil profile control the further spread or deposition of metals. The degree of human activities, mineral weatherability, and other pedogenesis-influencing processes control the mobility of metals in anthropogenic and naturally occurring metalliferous soils. Due to the variations in physico-chemical characteristics of metalliferous soils, metals can be more or less mobile to the other environmental compartments, especially biota. If not managed or treated adequately, those processes can influence food webs and food quality and pose further risks to human health. However, metalliferous soils are found to host specific plant communities consisting of species tolerant to the presence of metals and adopted to the unfavorable conditions of metal-rich sites. Those species become the base for development of particular remediation technologies, such as phytoremediation and phytomining. Applications of plant-based technologies for metal-rich soils were advanced upon discovering species with ability to hyperaccumulate metals. Additionally,

Metallophytes in Mineral Mining
DOI: https://doi.org/10.1016/B978-0-443-30110-0.00009-X

development of integration remediation approaches enabled further enhancing the effectivity of plants in environmental clean-up and overcoming of present shortcomings, especially longevity of phytoremediation process. Besides representing challenge in terms of overall environmental quality and health, metalliferous soils may also be observed as opportunities for additional obtaining of metal resources, particularly within the scope of environmental remediation and sustainable resource management.

2.2 Origin of metal-rich soils

Soils are composed of solid, liquid, and gaseous phases, mutually influenced by chemical and biological processes that determine their composition and function. Metal-rich or metalliferous soils contain elevated concentration of metals compared to ordinary soils. They originate from the weathering of naturally metal-enriched bedrocks or from contamination of soil by various industrial, agricultural, or urban processes. Particularly enriched metalliferous soils are those placed above or in the vicinity of metal ore deposits, which consist of valuable elements or minerals above background concentrations that can be economically extracted.

Natural metal-rich soils are of lithogenic origin, resulting from complex geogenic processes in the internal and surface layers of the Earth. Their formation and level of metal enrichment depends on geological, chemical, and physical processes involved in pedogenesis. Metal-rich soils can be derived from magmatic, metamorphic, and sedimentary bedrocks, where they reflect parent material's mineral composition to a greater or lower extent, depending on the involvement and strength of various pedogenic factors. The weathering of rock minerals is a significant natural contributor of metals to the soil system, and their levels in the soil are typically related to the elemental composition of the parent material.

Igneous rocks represent the source of many metal-rich soils over the world. Magma contains variety of chemical elements, and during process of magma differentiation (that involves cooling, partial melting, and crystallization on different temperatures) forms a various range of rocks with diverse chemical compositions. Igneous rocks are classified as intrusive when they form beneath the Earth's surface or extrusive when they form from magma that erupts onto and cools at the Earth's surface. Some of the most common intrusive magmatic igneous rocks are granite, basalt, gabbro, diabase, diorite, pegmatite, and peridotite. Although formed beneath the Earth's surface, igneous rocks can be uplifted and exposed to the surface conditions through tectonic orogenesis processes. Most common extrusive rock types are andesite, dacite, basalt, and rhyolite. Due to faster cooling time, they are usually formed of smaller size crystals in comparison to intrusive rock types.

During the process of magma cooling and solidification, particular metals become fractionated and concentrated in crystal lattice of different primary minerals. Metals can substitute other elements during the crystallization in process called isomorphic substitution. Another type of metal accumulation originates from circulation of hydrothermal fluids (heated solutions originating from cooling of magmatic bodies) through magmatic rocks in the Earth's crust, leading to the formation of hydrothermal ore deposits. Hydrothermal deposits usually include enrichment of elements such a copper, gold, lead, zinc, and molybdenum (Grosche et al., 2023).

Some of the most represented metals in metalliferous soils associated with magmatic rocks are nickel, copper, zinc, and chromium (Larocque and Rasmussen, 1998). Different types of igneous rocks, such as peridotite, basalt, and gabbro represent the sources of Ni, whereas silicate minerals such as olivine, hornblende, magnetite, and biotite can also be enriched with this element (Gonnelli and Renella, 2012). Concentration of Ni in ultramafic rocks can range up to 2000 mg/kg (Kabata-Pendias, 2011). Due to its siderophillic nature, Ni forms associations with sulfur, and it is present in various magmatic sulfide deposits around the globe.

Average concentration of copper in rocks is to be the highest in mafic igneous rocks, ranging 10–120 mg kg^{-1} (Kabata-Pendias, 2011). Majority of copper deposits are found in magmatic-hydrothermal porphyry deposits related to igneous rocks, followed by gold, silver, and molybdenum. Due to its affinity for sulfur, most abundant and economically important copper mineral is chalcopyrite ($CuFeS_2$). Zinc is element present in various igneous rocks, particularly basaltic (average 105 mg kg^{-1}) and granitic (up to 60 mg kg^{-1}), according to Mielke (1979). Most economically important zinc mineral is sphalerite (ZnS), commonly found in volcanogenic sulfide deposits. Chromium is also abundantly present in some igneous rocks (up to 3000 mg kg^{-1} in ultramafic rocks, according to Kabata-Pendias, 2011), while chromite mineral ($FeCr_2O_4$) typically found in mafic and ultramafics represents primary source of chromium extraction.

Sedimentary rocks are formed by decomposition, transportation, and sedimentation of preexisting rocks. They are characterized by very variable composition originating from diverse source of deposited material and sedimentation environments. Based on its formation conditions, sedimentary rocks are categorized into clastic sedimentary rocks, chemical sedimentary rocks, and organogenic (biogenic) sedimentary rocks. Clastic sedimentary rocks are composed of fragments and debris of previously existing rock types, ranging in grain size. Main clastic rock types are conglomerates and breccias, sandstones, siltstones, and claystones. Chemical sedimentary rocks are formed by minerals precipitation and evaporation of chemically enriched water, and are represented by evaporites, chert, limestones, and dolomite rocks. Biochemical and organic sedimentary rocks are formed from the remains of living organisms or from their shells, and the representatives are coal and certain types of limestones.

Sedimentary rocks can host metal deposits formed by the accumulation of metal-bearing minerals due to erosion, weathering, transportation, and deposition processes. Deposits formed by chemical sedimentation of iron minerals, called banded iron formations are one of the most significant source of iron ore worldwide (Hagemann et al., 2016). Sedimentary copper deposits, occurring as native copper, copper sulfides, or oxidized copper deposits, account as the world's second largest source of copper, accompanied by increased level of cobalt and silver (Squire and Keays, 2024). Black shale is fine-grained sedimentary rock containing >0.5% of organic carbon, and it is often hosting polymetallic deposits enriched with elements such as copper, nickel, zinc, molybdenum, vanadium, uranium, and elements of the platinum group and rare earth elements (REEs) (Parviainen and Loukola-Ruskeeniemi, 2019).

Metamorphic rocks exhibit changes in physical and chemical characteristics during process of metamorphism (under high pressure or increased temperature) within the Earth's crust. This process modifies the original rock (protolith) to create altered mineral composition and influence the formation of new minerals. Chemical composition of metamorphic rocks, including content of metals, usually remains similar to one of the protolith, although often altered and rearranged into new mineral formations. Based on the size of the area affected by metamorphism, two main types of metamorphisms may be distinguished: regional and local. Regional metamorphism occurs in the wide areas of the crust under heat and pressure that alters protolith, whereas local metamorphism occurs in much smaller areas when rocks are heated and modified by magma. Additionally, depending on the presence or absence of differential pressure and formation of characteristic layered mineral patterns, metamorphic rocks can be foliated or nonfoliated. Typical foliated metamorphic rocks are shale, phyllite, granite, schist, and gneiss, while characteristic nonfoliated (massive) metamorphic rocks are marble, quartzite, and serpentinite.

Processes present in metamorphic rocks, such as recrystallization, chemical mobilization by active fluids, and increased pressure and temperature could cause metal transport and their concentration into specific zones, as well as the formation of economically important metal deposits. This can happen during the process of metasomatism, contact, or regional metamorphism. Some of the best known deposits formed during metamorphism are gold deposits, skarns, and graphite deposits that include elements such as gold, copper, zinc, lead, and iron.

Gold deposits are often formed in relation to hydrothermal activities and orogenic processes, in host rocks such as gneiss, schist, and amphibolite (Groves et al., 2003). Skarn deposits form as a result of the replacement, alteration, and contact metasomatism by ore-bearing hydrothermal fluids that are found next to the igneous intrusions into sedimentary strata with carbonate-rich rocks. Skarns often host metal deposits such as iron (in magnetite), copper (in chalcopyrite, bornite, and covellite; Fig. 2.1), gold, and zinc-lead (sphalerite and galena; Fig. 2.2) (Mishra, 2000). Graphite deposits in metamorphic rocks, although not metallic, can sometimes be accompanied with gold, copper and zinc-bearing minerals.

Figure 2.1 Fragments of chalcopirite ($CuFeS_2$) dispersed in skarn.

Figure 2.2 Sphalerite (ZnS) in skarn.

Various igneous, sedimentary, and metamorphic rocks represent parent material for the formation of different soil types. This material, often referred as subsolum, is a starting material for soil formation and in has been classified by Juilleret et al. (2016) in four reference groups by taking into account their various weathering stages: regolith, saprolite, saprock, and bedrock. While bedrock refers to solid unweathered rock representing a base for soil formation, regolith presents unweather and unconsolidated deposited material that can be found above the bedrock surface. On the other hand, saprock refers to the in situ subsolum material that contains less than 20% of weathered minerals, while saprolite represents in situ chemically weathered geological material that retained some structure of its lithic component and contains more than 20% of weathered minerals (Graham et al., 2010).

The process of soil formation begins when loose parent material settles on a stable terrain or when bedrock becomes exposed at the surface of the Earth, and process of pedogenesis starts under the influence of other pedogenetic factors: topography, climate, living organisms, and time. During this process, physical, chemical, and biological weathering of minerals takes place. Along with weathering process, primary minerals from bedrock become altered and changed to secondary minerals, particularly clay minerals and iron and aluminum oxy-hydroxides. Weathering rate is governed by mineralogical composition of bedrock, referring to the presence of highly weatherable (calcite, olivine, pyroxene, amphibole, and biotite), moderately weatherable (muscovite and Kfeldspar), or minerals resistant to weathering (quartz and hematite) (Bonnet et al., 2022). The weathering products from rocks and minerals containing elevated metal concentrations further experience various pathways: they are transported by flowing water, bind with organic materials or mineral surfaces, integrated into secondary minerals, or absorbed by living organisms. Natural concentration of elements in soil that resulted from the lithologic and geochemical component and not influenced by antropogenic cause is referred as background value and is unique for each region or area (Bini et al., 2011), being naturally higher in case of metalliferous soils. Particular type of metal-enriched soils, formed in tropical and subtropical areas with hot and wet climate as a consequence of intense weathering of iron- and aluminum-rich parent materials, are laterite soils. They are enriched in elements such as iron, aluminum, and manganese, rich in clay and often of characteristic red color.

Besides soils naturally containing elevated concentration of metals, there are many areas of anthropogenically metal-enriched soils worldwide. They are mainly associated with human activities such as agriculture, urbanization, industrial processes, mineral exploitation and processing, or waste disposal. The type of prevailing activity determines which metals will accumulate in the case of anthropogenically contaminated soils.

Mining activities represent one of the most significant sources of metal pollution worldwide due to the ongoing mining operations and legacy of contaminated sites left after the completion of mining activities. Some of the most important environmental problems are produced during mining of gold, especially during the processing of arsenopyrite in sulfide hydrothermal gold deposits and gold amalgamation process with mercury (Alencar Meira da Silva and Davée Guimarães, 2024; Han et al., 2023). Comprehensive study of worldwide soils near gold mines conducted by Hou et al. (2023) revealed that they exhibit high contamination levels of As, Cd, and Hg that pose ecological and health risks to their surroundings. Similarly, analysis of soils in the vicinity of 102 copper mines around the world showed increased concentrations of Cu, Cd, and As and associated health risks (Chen et al., 2022a). An analysis of soils near Pb/Zn mines around the world showed increased concentrations of Pb, Zn, Cd, As, and Cu (Rouhani et al., 2025). Pollution assessment of soils near uranium mines showed increased concentrations of U, Cd, and As (Chen et al., 2022b), whereas soils near coal mines exhibited increased concentrations of Ni, Cd, Zn, and Hg (Xiao et al., 2020). Additionally, ancient mining and metallurgy sites worldwide left increased level of Ag, Cu, Hg, Pb, and Zn in surrounding soils (compared to geological background levels), influencing biogeochemical cycles and becoming incorporated into different components of surrounding ecosystems to this day (Asare and Afriyie, 2021).

Industrial activities are a notable contributor to anthropogenic soil pollution, causing the accumulation of different potentially toxic elements (PTEs) along the soil profiles near industrial sites and facilities. Kabir et al. (2012) noted that elements in soils such as As, Cu, Ni, Pb, and Zn were mostly associated with smelting processes and metal industries, while Mn, Cd, and Cr were recorded in soils impacted by textile and leather industry. Waste disposal and agriculture are some of the major causes of contamination on local level. Disposal of wastes represent significant contributor to the overall pollution especially in developing countries, generating increased levels of Pb, Cd, Cr, and Hg (Cachada et al., 2018). Mean values of As, Cd, and Hg from agricultural soils worldwide showed two- and threefold increase in comparison to China's and European guidelines, as well as the continental crust content (Nde et al., 2024). Urban development, follow by increasing rate of anthropogenic activities, is also one of the causes of soil metal pollution. Urban soils from 174 cities around the world were analyzed for the presence of metals, revealing the average concentration of As, Cd, Co, Cu, Hg, Ni, Pb, and Zn above the Earth's crust level, whilst exhibiting highest contamination factors for Cd and Pb (Adewumi and Ogundele, 2024).

2.3 Categories of metal-rich soils

Metalliferous sites can be distinguished based on the mineral composition of the substrate or the source of the metal pollution. According to the origin of metal enrichment, identified three categories of metalliferous sites: primary, secondary, and tertiary.

2.3.1 Primary sites

Primary sites have developed spontaneously and without human influence as a result of metal ore outcropping or natural weathering processes that contribute to the genesis of metalliferous soils. They are nowadays considered to be of a relatively rare occurrence, occupying small and patchy areas. Parts of ultramafic (sometimes called "serpentine") and calamine soils (formed of rocks rich in "calamine," i.e., zinc ore) that remained free of anthropogenic influence belong to such type of metalliferous sites. However, majority of sites with natural ore outcrop have been exploited by ancient or modern mining activities. Some high-altitude ultramafic sites in Alps or noninhabited ultramafic sites in Balkan Peninsula (Figs. 2.3 and 2.4) are representatives of such primary sites (D'Amico et al., 2009; Stevanovic et al., 2003).

Parent materials like peridotite or serpentinite are usually enriched with metals like Ni, Cr, and Co and poor in elements like Ca, P, and K, directly influencing their content in above-ground soils (Kierczak et al., 2021). There are, however, differences in physico-chemical characteristics of soils derived from various ultramafic rocks with different degrees of serpentinization (Marescotti et al., 2019; Pędziwiatr et al., 2018) found that degree of serpentinization regulates mineralogy and geochemistry of such soils, as well as the distribution and mobility of Cr, Ni, and Co. Primary minerals, present in different percentages in partially serpentinized lherzolites, massive serpentinites, and foliated serpentinites (72%, 65.5%, and 54% in the hummus accumulative soil horizon, respectively) were the main carriers of those metals. Researches (Kierczak et al., 2021) pointed to the higher phytoavailability of Ni on soils originated from serpentinized peridotite in

Figure 2.3 Ultramafic habitat on Mount Maljen, west Serbia.

Figure 2.4 Ultramafic habitat on Pešter Plateau, southwest Serbia.

comparison to those derived from serpentinite, as the olivines (major source of Ni in peridotites) are unstable in surface conditions in comparison to serpentines (major source of Ni in serpentinites). Pędziwiatr et al. (2018) supported this conclusion but found the opposite content of Ni in plants inhabiting these soils due to abundance of Ca and Mg on serpentinite sites and their interference with metal uptake. Calamine soils of natural origin are nowadays very rare and occur due to weathering of rock outcrops containing zinc- and lead-bearing minerals (such as sphalerite and galena). They are characterized by increase contents of Zn, Pb, and geochemically associated Cd (Wójcik et al., 2017).

2.3.2 Secondary sites

Secondary metalliferous sites are generated by anthropogenic activities, often resulting from mining processes and deposition of mine wastes. Many secondary sites have been developed on the locality of former primary metalliferous sites that were later modified and transformed by human activities (Baker et al., 2010).

Global land surface occupied by mining infrastructure and facilities like waste rock dumps, pits, water ponds, tailings dams, and processing plants is estimated to range between 57,000–66,000 km^2, based on the use of satellite data with limited precision (Maus et al., 2020; Tang and Werner, 2023; Werner et al., 2020) that mapped 295 metal mines and accompanying facilities around the globe that produce copper, lead-zinc, nickel, gold, silver, uranium, and platinum covering an area of approximately 3600 km^2. However, these values are under-valued due to lacking of data concerning artisanal, small-scale and illegal mining sites. Influence of mining sites on the diverse components of the environment is targeting much wider area, estimated to reach even up to 50 million km^2 of terrestrial land surface, if the impact of each mine is assumed to extend 50 km in radius (Sonter et al., 2020).

Many metal-enriched sites generated by mining activities or remaining after the closure of mine facilities are nowadays hosting miscellaneous vegetation, including various metallophyte plant species (Jakovljević et al., 2020; Ranđelović and Jovanović, 2023; Ranđelović et al., 2016). These soils significantly differ not only according to the present metal type, geochemical association of chemical elements, and their mobility in soil but also according to the soil texture, pH, organic matter content, and other physico-chemical characteristics, offering unique conditions for the development of metallophyte species. Majority of calamine and parts of ultramafic sites originating from primary mineralized outcrops that are nowadays modified by historical or present human activities also belong to the category of secondary metalliferous sites. The content of metals in such soils is increased due to their combined lithological and anthropogenic origin (Figs. 2.5 and 2.6). Table 2.1 presents some data on total content of metal(oid)s in secondary metallogenic sites around the world that are

Figure 2.5 Vegetation over serpentinites at abandoned asbestos mine Korlaće, Serbia.

Figure 2.6 Vegetation development in Piskanja coal washing basin, Serbia.

hosting different types of vegetation. However, total content of metals in soil does not fully reflect the true amount of those elements available to plant organism. Certain fractions of total element concentration that are more mobile in environmental conditions are considered available to plants. Fraction of metal bound to crystal aluminosilicate lattice is nonmobile in soils and therefore cannot be uptaken by living organisms. Other geochemical forms of elements such as water-soluble, easily exchangeable, and bound to carbonates are regarded as readily available to plants, while those bounded to Fe/Mn oxides and organic matter are considered to be only partially bioavailable (Rodríguez et al., 2009). Additionally, elements of anthropogenic origin are often found to be bounded to more mobile fractions in soil and are considered more available to plants in comparison to those of litogenic origin (Ondrasek et al., 2025).

2.3.3 Tertiary sites

The tertiary metalliferous sites are formed by the contribution of metals to the environment previously free of metals due to the processes such as metal-enriched sediment deposition or metal-polluted air deposition from industrial facilities, municipal waste disposal, or agricultural activities including use of pesticides and fertilizers that contain metals (Baker et al., 2010). During these processes metals are transported by different environmental media to a shorter or longer distances, potentially influencing wider area. Some of the most highlighted examples of tertiary metalliferous sites

Table 2.1 Secondary and tertiary metalliferous sites with determined total metal content.

Locality	Type	Cu[a]	Ni[b]	Zn[c]	Pb[d]	Cd[e]	As[f]	Co[g]	Sb[h]	Tl[i]	Method[j]	References
Sarcheshmeh Cu mining area, Iran	Secondary	1330	34	1500	770						HCl/HNO_3	Ghaderian and Ghotbi Ravandi, 2012)
Miguel Vacas abandoned Cu mine, Portugal	Secondary	3480	141	118	63		950				$HNO_3/HF/H_2O_2$	Abreu et al. (2008)
Caporciano Cu mine dump field site, Italy	Secondary	7 302	115	765	20	3.4	2				$H_2O/HF/HClO_4/HNO_3$	Andráš et al. (2021)
Limni Cu mine tailings, Cyprus	Secondary	1534	122	4132	28.6	6.4					HNO_3/H_2O	Christou et al. (2017)
Kapunda historical Cu mining site, Australia	Secondary	3822	25.3	27.5	31	0.2	8.3				HCl/HNO_3	Guo et al. (2009)
Hengyang City Pb/Zn mine, Hunan Province, China	Secondary			651	947	5.7					$H_2O/HF/HClO_4/HNO_3$	Cao et al. (2022)
Rubiais abandoned Pb/Zn mine, Spain	Secondary	160	41	52000	6100	95					HCl/HNO_3	Monterroso et al. (2014)
Boleslaw calamine waste heap, Poland	Secondary		36	40000	1650	170					$HClO_4$	Olko et al. (2008)
Koshk Pb/Zn mining area, Iran	Secondary			3442	2831						$ClO_4/HNO_3/HCl$	Mahdavian et al. (2017)
Barbadalhos Pb mine, Portugal	Secondary	193	51	517	9330		208				$HF/HClO_4/HNO_3$	Pratas et al. (2013)
Penas Albas serpentinite, quarry, Spain	Secondary	145	2040	34							$Li_2B4O_7–LiBO_2/HCl/HNO_3$	Lago-Vila et al. (2015)
Lubumbashi Cu/Co metalliferous soil anthropogenically modified, Congo	Secondary	9720		622				335			$HF/HClO_4$	Pourret et al. (2015)
Abandoned Pb/Zn Ingurtosu Mine tailing site, Italy	Secondary			7400	1800	56					$HNO_3/HClO_4$	Barbafieri et al. (2011)
Abandoned Allchar As/Tl mine site, North Macedonia	Secondary						76100		345	4280	HCl/HNO_3	Jakovljević et al. (2024)

(*Continued*)

Table 2.1 (Continued)

Locality	Type	Cu[a]	Ni[b]	Zn[c]	Pb[d]	Cd[e]	As[f]	Co[g]	Sb[h]	Tl[i]	Method[j]	References
Stolice Sb mine tailing dam spill (soil/ sediment), Serbia	Tertiary			626	188	5.2	1542		590		HCl/HNO_3	Ranđelović et al. (2019)
Soil in vicinity of Lastenia metal foundry smelter, Argentina	Tertiary		63	1640	8714				28.6		$HF/HClO_4/HNO_3$	Fernandez-Turiel et al. (2001)
Soils around Pb/Zn smelter in Huixian, China	Tertiary	52.6		165	4090	26.1					$H_2O/HF/HCl/HNO_3$	Zhan et al. (2014)
Soils of former e-waste incineration site, Guangdong province, China	Tertiary	21400	132	8970	7720	46.8					HNO_3/ $HClO_4$	Luo et al. (2011)
Agricultural soils of Argolida basin, Greece	Tertiary	74.7	147	74.9	19.7	0.54	6.9	22			$HNO_3/HCl/H_2O$	Kelepertzis (2014)
Agricultural soils of Sinú River basin, Colombia	Tertiary	2522	1391	2632	0.13	0.08					HCl/HNO_3	Marrugo-Negrete et al. (2017)
World-soil average		38.9	29	70	27	0.41	6.8	11.3	0.67	0.5		Kabata-Pendias and Pendias (2011)

[a] Mean or maximum (if available) obtained values in various types of metalliferous soils under vegetation are given in mg kg^{-1}.
[b] Mean or maximum (if available) obtained values in various types of metalliferous soils under vegetation are given in mg kg^{-1}.
[c] Mean or maximum (if available) obtained values in various types of metalliferous soils under vegetation are given in mg kg^{-1}.
[d] Mean or maximum (if available) obtained values in various types of metalliferous soils under vegetation are given in mg kg^{-1}.
[e] Mean or maximum (if available) obtained values in various types of metalliferous soils under vegetation are given in mg kg^{-1}.
[f] Mean or maximum (if available) obtained values in various types of metalliferous soils under vegetation are given in mg kg^{-1}.
[g] Mean or maximum (if available) obtained values in various types of metalliferous soils under vegetation are given in mg kg^{-1}.
[h] Mean or maximum (if available) obtained values in various types of metalliferous soils under vegetation are given in mg kg^{-1}.
[i] Mean or maximum (if available) obtained values in various types of metalliferous soils under vegetation are given in mg kg^{-1}.
[j] Method for determination of total element concentration is presented.

Figure 2.7 Primary colonization of Stolice antimony mine flotation tailing spill, Serbia.

are those formed after environmental accidents, such as tailing dam spills that are considered to have severe influences on the environment. Such areas not only require thorough examination of short and long-term effects on the environmental quality of impacted areas can also serve as unique habitats for the development of various metallophyte species that can help to mitigate immediate consequences (Fig. 2.7) (Ranđelović et al., 2019).

Another representative example are soils in the vicinity of metal ore smelting sites, usually contaminated within certain radius from the smelter facility. Due to the way they are formed, tertiary metalliferous sites are characterized by increased presence of metals of anthropogenic origin that could be potentially more bioavailable to the plants and therefore additionally influence plant species' composition and chemistry.

2.4 Worldwide distribution of metal-rich soils

Metalliferous soils are non- uniformly distributed throughout the world. Their genesis is connected with the presence of natural mineralization in the Earth`s crust and with the different types and intensities of anthropogenic activities all over the globe. Mineral deposits represent natural accumulation of metal-bearing minerals in greater concentration than in the average Earth's crust lead by geological processes. Metallic mineral deposits are concentrated in certain areas of the Earth's crust in some cases reaching the Earth's surface. If mineral deposit has such abundance of metal-bearing minerals that it can be economically profitable for extraction, it is called ore deposit. Such features of the lithosphere are causing the increased metal content during formation of various types of soils, usually gathered by the common name "metalliferous." Background values of metals in such soils are often increased compared to the average levels commonly find in soils of the world.

Once exposed to the surface conditions, processes of mineral weathering, pedogenesis, metal distribution, speciation, and mobility toward environment and biota start taking place. Metalliferous soils are known to host specific vegetation comprising of plants able to tolerate increased levels of metals, called "metallophytae." Some of the most prominent examples of sites suitable for development of metallophytae are ultramafic and calamine soils.

Global distribution of ultramafic soils is tightly intertwined with the distribution of ultramafic rocks. Ultramafic and mafic sites have a notable role in metallogeny, as their unique composition and formation processes often lead to genesis of economically significant metal deposits of Ni. Two forms of Ni deposits associated with ultramafic and mafic sites can be distinguished—sulfide and laterite deposits (Fig. 2.8). Although around

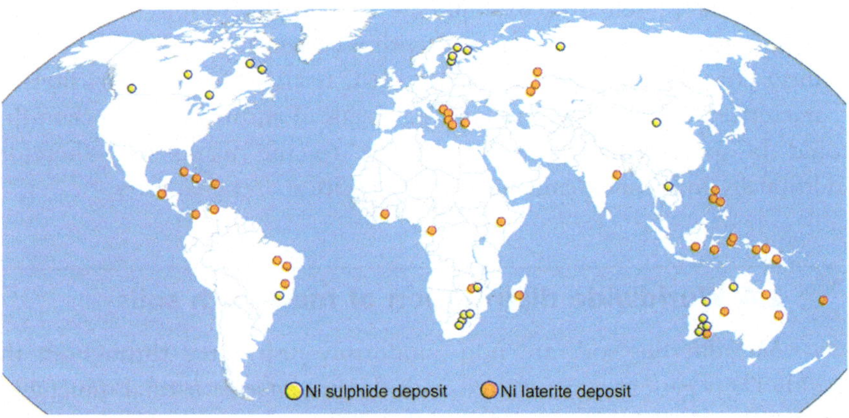

Figure 2.8 Global distribution of massive Ni sulfide and laterite deposits. Map lines delineate study areas and do not necessarily depict accepted national boundaries. Modified with permission from Hoatson, Jaireth, and Jaques, 2006, and Berger, V.I., Singer, D.A., Bliss, J.D., Moring, B.C., 2011. Ni-Co laterite deposits of the world – Database and Gradde and Tonnage Models. Open-File Report 2011-1058, US Geological Survey.

60% of world Ni resources have been found in laterites, due to the complex extraction treatment, majority of world Ni (over 60%) is actually acquired from sulfide ore deposits (Mudd, 2010).

Nickel sulfide deposits are formed in association with magmatic processes, where Ni is usually geochemically associated with Cu but also with Au, Ag, and platinum group of metals. Some of the most prominent Ni sulfide deposits in terms of ore richness and area occupied in the world are Sudbury Basin (Canada), Norilsk-Talnakh (Russia), Pechenga (Russia), Jinchuan (China), Bushveld Complex (South Africa), and Kambalda (Australia).

Nickel laterite deposits are formed by intense weathering of ultramafic rocks in hot and humid climate, in the 23.5 parallels of the north and south of the equator (Bustamante-Rúa et al., 2020). In such settings, Ni enrichment is usually accompanied with Co. Process of lateritization leads to enhanced concentration of Ni and Co up to 30 times in comparison to their content in the bedrock (Elias, 2002). Presence of nickel lateritic deposits is associated with ultramafic rocks in areas of tropical climate. Some of the richest Ni deposits in the world are concentrated in Indonesia (accounting for 26% of global mining production, according to Soh Tamehe et al. (2024)), Philippines, and New Caledonia, where they are actively exploited. Significant deposit of Ni laterite ore is also found in Sukinda, Indonesia. Besides owning significant world reserves of nickel, the region of South and Southeast Asia pose the largest global surface exposures of ultramafic rocks in the world that represent unique habitats for the development of metallophyte and hyperaccumulator plant species and one of their global hotspots (Galey et al., 2017). Laterite Ni deposits of relatively low economic grade have also been exploited within the Yilgarn craton in Western Australia (Gole and Barnes, 2020), and the area is recognized as rich in plant diversity and specialist taxa (Gibson et al., 2012). Similarly, Ni laterites of South America and Caribbean (Gutemala, Costa Rica, Venezuela, Cuba, Colombia, and Brazil) also host some of the world most significant Ni deposits and ultramafic outcrop surfaces that are rich in endemic and hyperaccumulating species (around 130 Ni hyperaccumulator species have been registered in ultramafic flora of Cuba, which is the largest concentration of hyperaccumulators in the world, according to (Reeves et al., 1999). The most significant Ni laterite deposits in Africa are found in equatorial range of Burundi and Côte d'Ivoire but also in West Africa (Musongati) and Madagascar. Regions of Ural in Russia and Aktobe in Kazakhstan are also characterized by rich Ni laterite deposits. Significant deposits of Ni laterites are placed in southern eastern parts of Europe. Laterite deposits of nickel in Balkan Peninsula and Turkey were formed in restricted time of warm and humid conditions, after which they have been set in temperate and arid climates (Soh Tamehe et al., 2024). Those laterites were formed in the Early Cretaceous and were prone to the complex formation processes, including partial preservation *in situ* or transport and redeposition in shallow marine environments with repeated regressions, resulting in alternation with sedimentary rocks or metamorphosis to varying degrees due to the regional tectonic activity (Eliopoulos et al., 2012). Consequentially, the largest ultramafic outcrop surfaces in Europe are found in the Balkan Peninsula, covering parts of Bosnia, Serbia, Bulgaria, Albania, Greece, and Turkey. This region comprises different types of ultramafic soils whose flora is characterized by high level of endemism (335 taxa, according to Stevanovic et al. (2003)). Increased level of Ni in ultramafic-derived soils enabled development of metal-tolerant species, resulting in 27 Ni hyperaccumulating taxa recorded on the Balkan Peninsula (Jakovljevic et al., 2022).

Another type of metalliferous soils is related to the distribution of rocks containing zinc ores (most often associated with Pb and Cd) that occur naturally or are linked to historical or present mining activities. Generally, Pb/Zn deposits could be classified as volcanogenic sulfide deposits or non-sulfide deposits, derived from the weathering and oxidation of deposits that are often often carbonate-hosted (Wen et al., 2016). Some of the world most prominent sulfide Zn/Pb deposits are placed in the USA (Red Dog Mine in Alaska), Australia (Mount Isa, Golden Grove, Broken Hill), China (Jinding, Tianbaoshan, Huize), Peru (Tambogrande, Perubar), India (Rampura-Agucha), and Iran (Gol-e-Zard, Koushk) (Huston et al., 2006; Hou et al., 2023; Lewchuk et al., 2004; Mishra and Bernhardt, 2009; Rajabi et al., 2020; Zarasvandi et al., 2021; Zhou et al., 2023). Notable non-sulfide Zn/Pb deposits are found in China (Huoshaoyun), Brazil (Vazante), Namibia (Skorpion), Yemen (Yemen), Poland (Silesia--Cracow), Belgium (La Calamine), and Italy (Iglesias) (Borg et al., 2003; Chen et al., 2023; Coppola et al., 2008, 2009; Monteiro et al., 2007). Calamine- or Pb/Zn(Cd)-rich soils are closely associated with Pb/Zn non-sulfide mineral deposits hosted in carbonates and their extraction activities (Boni and Mondillo, 2015). Term "calamine" originates from the old name for secondary zinc minerals having different meanings in English (smithsonite, $ZnCO_3$) and American (hemimorphite, $Zn_4(Si_2O_7)(OH)2{\cdot}H_2O$) use, although nowadays this term is rarely used due to ambiguity. Typical calamine mineralization has been found in old non-sulfide deposits of Europe (in Belgium, Italy, and Poland), resulting in the development of Zn-, Pb-, and Cd-rich calamine soils that support diversity of metallophytae species (Wójcik et al., 2017).

Metalliferous soils linked to higher copper levels are primarily in areas where copper mineralization occurs naturally or where copper mining is practiced. Copper deposits are found worldwide, with certain regions being particularly rich in reserves. Based on their geological formation, several types of copper deposits may be singled out. Porphyry copper deposits are most significant sources of copper on global level, formed in association with intrusive igneous rocks. They are usually of low grade but massive, and their ore exploitation is usually done by surface mining. These deposits are typically places along tectonic plate boundaries, especially in regions associated with current or paleo subduction zones such as Circum-Pacific belt, Tethys belt, and Central Asia belt (Liu et al., 2020; Park et al., 2021). Largest copper deposits in the world are placed in the Atacama region of Chile, the leading copper-producing country in the world. Significant porphyry copper deposits can also be found in Balkan and east Turkey, eastern parts of Australia, northwest China and Tibet Plateau, Ural Mountains, and eastern Russia (Sun et al., 2015).

Sedimentary copper deposits are the second most important type of global copper deposits, formed in sedimentary rocks via chemical processes regularly associated with ancient marine or lacustrine environments. The largest sedimentary copper deposit in the world is Central African Copperbelt on the territory of Democratic Republic of Congo and Zambia, known as second largest source of copper and the largest source of cobalt globally (Singer, 2017). Soils with natural Cu–Co enrichment are formed by weathering of rock outcrops with high concentrations of those elements. Additionally, Cu–Co metalliferous sites are abundant in the African Copperbelt area due to the presence of many mine waste and excavation sites. Soil Cu and Co concentrations primarily influence plant species richness and level of endemism in those areas (Mukube et al., 2024). Presence of 12 hyperaccumulators of copper and 15 hyperaccumulators of cobalt was recorded on the territory of Copperbelt in Zambia (Brooks et al., 1980), while several geobotanical indicators for prospection of Cu deposits were recorded in Congo (Wójcik et al., 2017).

Volcanogenic massive sulfide deposits are formed by hydrothermal activities in submarine volcanic environments. They are rich in copper sulfides often, along with zinc, lead, gold, and silver. Some of the most important volcanogenic massive sulfide deposits in the world are in Canada (Kidd Creek, Abitibi, Bathurst, Flin Flon, and Buchans), Sweden (Skellefte district, Bergslagen district, and Caledonian belt), Spain and Portugal (Iberian Pyrite Belt), Russia (Urals), and Australia (Yilgarn, Pilbara, and Mount Morgan) (Allen and Weihed, 2002).

Iron oxide–copper–gold deposits are associated with iron oxide minerals and formed via magmatic-hydrothermal processes. Prominent global deposits of this type are in Australia (Olympic Dam and, Prominent Hill), Chile (Candelaria), and Brasil (Salobo and Sossego) (del Real et al., 2023). Copper skarn deposits are formed through chemical and physical alterations of minerals during contact of intrusive igneous rocks with carbonate-rich rocks (Fig. 2.1). They are found along subduction zones and continental margins. Largest copper skarn deposit in the world is placed in Peru (Antamina in Andes), while there are also notable deposits in Papua New Guinea (Ok Tedi mine on Star Mountains), Indonesia (Ertsberg), and China (Jiama) (Misra, 2000).

Thallium-rich soils originate from natural sources (ore deposits and volcanoes) or anthropogenic contamination (coal combustion and metal smelting) (Karbowska, 2016). Average concentration of thallium has been find to be highest in acid rocks (up to 2.3 mg kg^{-1} in granites and gneiss) and sedimentary rocks with increased content of clay (Kabata-Pendias, 2011). Thallium is highly toxic element, readily available for plants and humans and known to easily enter food chain (Xiao et al., 2004). Some of the richest thallium deposits in the world are discovered in Russia (Norilsk and Beshtau), North America (Getechll, Garlin, Franklin, and Rambler), China (Lanmuchang, Yunfu, Yunnan Lanping, and Anhui Xiangquan), and North Macedonia (Allchar and Lojane) (Liu et al., 2019). Hyper-accumulation of thallium by species of genus *Viola* has been observed in abandoned Allchar mine deposit in north Macedonia (Jakovljević et al., 2023).

Deposits of critical REEs have limited global occurrence confined to several countries: China (that dominates world REE production with 70% share, including deposits such as Bayan Obo, Maoniuping, Zudong, and Longnan), Russia (Khibiny and Lovozero), Brazil (Caldeira and Pitinga), the USA (Bear Lodge and Mountain Pass), Australia (Mountain Weld), Canada (Thor Lake and Kipawa), and South Africa (Songwe Hill) (Borst et al., 2020; Chen et al., 2024). The REEs are contained and currently exploited from hard-rock sources (carbonatite and alkaline rocks) and from weathered crusts at the Earth's surface (ionic clays and placer deposits) (Liu et al., 2023). Importance of REEs for world industry, particularly high-technology sector, is continually growing, aiming for technologies of REE extraction from secondary sources and enriched soils. Additionally, there is a growing interest in search for REE hyperacummulators over the globe (Purwadi et al., 2025).

Additionally, metalliferous soils can form in the metal-free environment due to the some sort of pollution and metal transport. Industrial activities, environmental accidents, and agricultural activities are some of the main factors influencing their placement and size. Some of the most severe environmental accidents are considered to be mine tailing dam failures. They often result in significant contamination of watercourses and soils over large areas, destruction of infrastructure, and loss of human lives. Highest frequency of tailing dam failures for the period 1915–2021 is recorded in America (107) and Asia, particularly China (43) (Wang et al., 2022; Lin et al., 2022). Majority of failures were associated with processing of metal ores, namely, copper, gold, iron, zinc, and lead, influencing the type of pollution that occurred. Major environmental disaster of such type in terms of magnitude happened in 2015, when iron mine tailing dam in Mariana, Brazil, collapsed releasing 43 million m^3 of iron ore tailings polluting 670 km of watercourses from the nearby Doce River to the Atlantic Ocean (Carmo et al., 2017). During this accident, around 1500 ha of natural vegetation and over 90% of the river riparian habitats were devastated (Couto et al., 2021). In 1998. mine tailing dam failure in Aznalcóllar, Spain, caused release of 2 million m^3 of pyrite sludge into nearby river, flooding around 4000 ha of surrounding soils with contaminated sediment containing excess Zn, Pb, Cd, and Cu (Galán et al., 2002). In 2014, abandoned antimony mine tailing dam was damaged by landslide causing release of 100,000 m^3 of tailings slurry into nearby Kostajnička river, contaminating 30 km of riverbed and 360 ha of agricultural soils with As, Cd, Pb, Sb, and Zn. Three years after the spill impacted soils were still excessing remediation values for As and Sb (Ranđelović et al., 2020).

Another source of great contribution to soil metal pollution are smelting activities. Hou et al. (2006) found that Pb isotope signature from Home copper smelter (Canada) could be traced more than 100 km downwind in soil organic horizons. McMartin et al. (1999) accessed dispersal patterns of metals from base metal smelter Flin Flon (Canada) in surface soils, finding that it averages 70 km for Cd, 76 for Pb, 90 km for Cu, and 104 km for As.

Intensive agricultural activities connected with enhanced use of pesticides may also elevate concentration of metals in soils. Around 242 million ha, accounting for approximately 16% of global cropland, is affected by increased level of at least one metal (Hou et al., 2025). Some of the most impacted areas are located in China, India, and Middle East. Certain combination of elements can be characteristic for agricultural land use of particular areas, or may identify the main chemicals used in agricultural and crop management. Meta-analysis of metal pollution of agricultural soils of China conducted by Huang et al. (2019) found higher concentration of Cd, Pb, Cu, and Zn in paddy and vegetable fields and generally higher concentration of metals in agricultural soils of southwest and south coastal areas of China. Accessing content of metals in European agricultural and grazing soils, Albanese et al. (2015) found highest Cu concentrations in agricultural soils of Italy, Greece, southern Balkans, and some provinces of France.

2.5 Remediation and management of metal-rich soils

Content of metals in soils is increasing on a global level in last decades, presenting significant environmental issue, influencing ecosystems and biota. Through the soil, metals enter the food chains or get transported to the water bodies. As they are non-biodegradable, they tend to accumulate in soil matrix, where they are chemically transformed. Metal speciation in soils depends on various conditions, influencing their overall mobility, bioavailability, and toxicity for organisms. Once accumulated in soils, metals can cause deterioration of soil quality and fertility and impair plant growth and soil microbial processes. Eventually, metals enter the food chains and affect human health.

Remediation and management of soils containing elevated levels of metals aiming at restoring soil to a healthy state or control the pollution are important for mitigating both the environmental and health impacts. This is particularly important in areas with history of mining, industrial pollution, or intensive agricultural activities. Soil remediation methods can be roughly divided to physical, chemical, and biological. Combined approaches that are integrating multiple methods are also emerging, aiming to enhance remediation efficiency. Type and level of contamination, metal speciation in soil, soil characteristics, and intended land use are some of the most important factors to determine the suitable remediation technology.

Physical remediation techniques focus on physical removing or separation of soil from the surrounding environment in order to reduce exposure to contaminants. It involves techniques such as soil excavation, replacement, or isolation. Excavation is a direct method of soil removal from the contaminated site for isolating it or handling on the other site. The process is considered effective for sites with concentrated and shallow contamination; however, it is often related with high costs especially when applied to the larger areas (Goldenberg and Reddy, 2014). The replacement approach is used to enhance the capacity of the soil and dilute pollution concentrations by replacing contaminated soil, either whole or partially, with fresh uncontaminated soil. This way, the amount of metal ions in the soil is significantly lowered and soil functional capacity is increased. Soil replacement is often used in small, polluted industrial areas due to the high application costs; however, more recently, methods for replacing soil with substrates that contain additions that improve remediation efficacy have been studied (Jiang et al., 2021). Soil isolation is remediation strategy that separates contaminants within the certain area by creating physical barriers in order to prevent them from dispersion and reduce risks to environment and biota. Several approaches can be used for isolation: capping (surface isolation with a material layer), vertical barriers (creation of cut-off walls), and encapsulation (complete isolation of a contaminant part with impermeable barrier). These methods are used in case that contaminated area is physically stable and immediate exposure needs to be minimized; however, the long-term stability of the immobilized metals needs to be carefully monitored.

Chemical remediation techniques focus on chemically transforming contaminants into less harmful state or their neutralization. Some of the most often applied techniques are chemical oxidation or reduction, soil washing with chemical agents (surfactants, flocculants, acids, etc.), chemical fixation using stabilization/solidification with chemical binders, and chemical precipitation. Using a range of extractants and chemicals that leach the metal ions is considered to be an effective way to treat the contaminated soil. A number of variables, including the target pollutants' characteristics, the extraction solution's bonding strength, and the surrounding environment, affect the soil washing procedure. The use of appropriate reagents and investigation of optimal conditions for the leaching of metal ions from contaminated soils have been readily investigated in previous years (Zheng et al., 2022). Recovery and treatment of washing agents is another important issue, as it significantly influence the cost-effectiveness of the process and therefore its practical application, while their improper treatment may additionally lead to the generation of secondary pollution (Zhang et al., 2022). Chemical fixation approach relies on immobilization of available metal phases into stable forms by adsorption, precipitation, or complexation mechanisms. Some of the amendments often applied for fixation purposes are minerals (lime, clay minerals, and phosphates) and industrial by-products (fly ashes, sludges, and red mud) (Lwin et al., 2018). While the metals are still present in soil in nonavailable forms, continual monitoring is needed to assure the efficient metal fixation over longer periods of time.

Biological remediation techniques use living organisms—microorganisms and plants—to remove or decompose soil contaminants during their metabolic processes. Microbial remediation is based on the ability of microorganisms to degrade or transform contaminants from the soil into nontoxic, less toxic, or neutral, more stable forms. During these processes, biosorption, enzymatic transformation, or metal precipitations are carried out. Some of the microbial species used in bioremediation of metals are *Flavobacterium* sp., *Pseudomonas* sp., *Bacillus* sp., *Azotobacter* sp., *Arthrobacter* sp., *Aspergilus* sp., and *Ganoderma* sp. (Verma and Kuila, 2019). The type of soil, physico-chemical characteristics, and degree of metal pollution at the contaminated site are some of the variables that influence the choice of living organism for the bioremediation process. Microbial remediation is considered a cost-effective, green solution that maintains the soil structure and evades secondary pollution at remediated sites (Verma et al., 2023). Biological remediation method that uses plants to reduce the concentration of pollutants in soil or water is called phytoremediation. During such remediation process several mechanisms can be applied, determining the type of phytoremediation: metal degradation (phytodegradation, by root metabolites, or plant enzymes), accumulation (phytoextraction and accumulation of metals in plant above-ground parts), evaporation (phytovolatilization and plant converse pollutants into gaseous state), and metal immobilization (phytostabilization and accumulation of metals in plant underground parts). Identification of most suitable plant species for soil phytoremediation has been the subject of extensive research in last decades. Use of metallophytes, group of plants that can tolerate increased levels of metals in soil, has been particularly encouraged (Matanzas et al., 2021; Ranđelović et al., 2024). Such species are naturally adapted to metal-rich soils and can be used in phytoremediation of anthropogenically polluted sites. A specific type of metallophytes, capable of accumulating extremely high concentration of metal(oid)s in their above-ground parts are called hyperaccumulators. More than 700 plant species are recognized for the ability to accumulate metals above the defined hyperaccumulation threshold (above 100 mg kg^{-1} Cd, Se, and Tl; above 300 mg kg^{-1} Co, Cu, and Cr; above 100 mg kg^{-1} Ni, As, and Pb; above 3000 mg kg^{-1} Zn; and above 10000 mg kg^{-1} Mg) in natural conditions so far (Reeves et al., 2018). Hyperaccumulators concentrate the metals in harvestable parts of their biomass, being able to gradually decontaminate the soil. However, many of hyperaccumulators are characterized by low biomass, which is often seen as an obstacle for the more efficient use in real field conditions. Application of technologies for enhancing biomass production of hyperaccumulating species and use of intercropping systems are some of the proposed directions for stringer integration of hyperaccumulating plants into phytoremediation programs (Álvarez-López et al., 2016; Kama et al., 2024).

Physical, chemical and biological soil remediation methods possess advantages and disadvantages (Table 2.2) that influence their practical implementation and frequency of application in remediation programs. In

Table 2.2 Pros and cons of physical, chemical, and biological methods of soil remediation.

Remediation method	Pros	Cons
Physical	– Fast and efficient – Suitable for diverse metals – Complete metal removal is possible	– Expensive and labor-intensive – Environmentally disruptive – Suitable for smaller areas
Chemical	– Fast – Effective for metal contaminants – Range of chemical agents available	– Generates waste – Potential secondary pollution – Often degrades soil quality – Can be expensive – Requires postremediation monitoring
Biological	– Environmental-friendly – - Cost-effective – Improves soil quality and health – Suitable for large medium-contaminated areas	– Slow – Sensitive to environmental conditions – Limited to certain contaminants – Can generate metal-enriched waste (phytoextraction)

order to reduce disadvantages, maximize effectiveness and reduce costs, integrated approaches to soil remediation are developed and applied. These approaches are intensively tested in recent years in order to optimize the methods and overcome the challenges of diverse site-specific contamination problems.

Integrated remediation approaches including phytoremediation have been particularly investigated in order to achieve synergy of environmentally friendly and time-efficient technologies. Assisted use of chelates, organic amendments, and microorganisms were some of the methods found to improve the overall effects of phytoremediation (Liu et al., 2024). Joint use of biodegradable chelating agents and acidophyllic bacteria showed synergistic effect on the removal of Cu from surface layers of Technosols formed on reclaimed flotation tailing site in Bor, Serbia, by removing 99.6% of bioavailable fraction of Cu (Ranđelović et al., 2015). Joint application of electrokinetic remediation and phytoremediation on lead-, arsenic-, and cesium-contaminated soils has showed significant increase of elements bioavailability under treatment with electrokinetic field due to the lowering of soil pH, which consequently lead to increased accumulation of elements in roots and shoots of *Brassica juncea*, Brassica *oleracea*, and *Spinacia oleracea* (Mao et al., 2016). Integrated phytoremediation system, consisting of selected rhizosphere microbes and metal-accumulating tree species *Acacia saligna* and *Eucalyptus camaldulensis* was applied to multicontaminated former Zn smelter industrial waste disposal by Guarino and Sciarrillo (2017). Inoculation with rhizosphere microorganisms lead to the increase of species biomass and improved phytoremediation capacity (up to 61% for total content of As, 56% for total content of Cd, 52% of total content of Pb, and 83% of total content of Zn). In Cd-contaminated soil, an integrated soil metal removal system consisting of the iron-biochar nanocomplex, the hyperaccumulator plant *Leersia hexandra*, and metal-resistant microbes community demonstrated effective metal removal (Wang et al., 2020). While iron-biochar nanocomplex decreased bioavailable fraction of Cd in soil, presence of resistant microbes impacted the uptake of Cd in above-ground parts of hyperaccumulator plant. Additionally, soil quality was improved, as it showed increase in overall microbial biomass, enzymatic activity, and content of organic matter. Design and application of integrated remediation methods still require investigation and optimization in terms of specific and real field site conditions, as well as the postremediation monitoring to prove efficiency and sustainability over time. Future of remediation technologies in the

21stcentury lies within development of more sustainable and environmentally friendly approaches, preferably within the circular economy approach.

In addition to active remediation, metal-rich soils need to be managed over longer period of time in order to effectively lower metal bioavailability, thus reducing risks of metal dispersion into environmental compartments, inclusion in food chains, and eventual influence on human health. Application of efficient management practices could lessen the detrimental effects of metal-rich soils on natural and agricultural ecosystems and urban and industrial environments. Addition of various amendments can help improve properties of metal-rich soils and reduce the bioavailability of metals. Liming (application of calcium-containing materials) is a commonly used and effective strategy to reduce the mobility and bioavailability of toxic metals in soils, most often in acidic environments. Incorporation of various organic materials (compost, biochar, or plant residues) is an effective environmentally friendly strategy to reduce the availability of heavy metals in metal-rich soils while additionally improving the soil health. For example, incorporation of wheat straw crop residues to moderately metal-contaminated rural and urban soils from northern and central Greece increased soil quality indices while simultaneously limited availability of toxic metals (Cd and Pb) and increased availability of trace elements (Zn anc Cu) (Golia, 2023). Addition of biochar to metal-rich soils also contributes to reduced metal bioavailability, through reactions such as ion exchange, complexation, precipitation, improvement of soil pH, and impact on soil cation exchange capacity and soil organic carbon content (Mansoor et al., 2021). Use of metal-accumulating or metal-tolerant plants and their cultivars for management of metal-containing agricultural soils and metal levels in crops is also recognized approach (Cao et al., 2021). Adoption of crop rotation practices has additionally been proved to manage Al, Cd, Cu, Ni, Pb, and Zn concentrations in wheat, particularly with grass/clover (Cooper et al., 2011). In cases that full remediation is not a viable option, other management activities including land use restrictions, restricted approach, and rising public awareness about the risks of metal-rich soils and associated environment can contribute to the protection of environment and human health.

2.6 Metal recovery as a resource from metal-rich soils

The growing need for metals in electronics, renewable energy, and other industries makes metal recovery from secondary sources an emerging field of study. Recovering metals from metal-rich soils is a novel approach that addresses contamination and transforms waste into valuable resources. Valuable metals are extracted from metal-enriched soils or mine tailings, which are frequently the consequence of past mining operations, pollution, or naturally occurring mineral deposits. In addition to addressing natural soil enrichment, metal recovery can improve soil quality and wider the spectrum of potential soil use for agricultural or other purposes. Application of recovery methods not only mitigates environmental risks but also contributes to the sustainable resource management. Development and improvement of such methods is of particular importance for countries having metal-rich soils and developed mine industries with historical mining heritage. Metal recovery from variety of metalliferous soils is often paired with remediation technologies as an upgrade so that beside land preservation or reclamation there is a possibility to obtain the metal resource by applying adequate technologies.

Leaching technology has been widely applied in the field of extractive metallurgy for mineral treatment in order to extract valuable metals. Metals are dissolved and separated from the solid matrix using a liquid solvent, usually water or acid. Same process is often used for removing metals from contaminated soils. This procedure can be utilized in both environmental clean-up initiatives (such remediating damaged soil) and in the extraction of metals for commercial reasons. However, recovery of metals from soil leaching process was generally investigated more scarcely. Additionally, soil organic matter content has been found to significantly influence the metal leaching rate and time. Ferraz and Lourençlo (2000) and Wu et al. (2020) studied the possibility of metal recovery from soil leaching process with the addition of sodium alginate–coated FeSSi nanocomposites. At the end of a leaching process, sodium alginate–coated FeSSi nanocomposites were easily separated from soil and use for metal recovery, and the effectiveness of metal recovery from soil and solution reached 63% and 99%, respectively. Treatment of calamine natural deposits with acidic solutions of HCl and HNO_3 resulted in the recovery of 39% of Zn (Swęd et al., 2023). Acid treatment of two contrasting ultramafic soils revealed that citric acid had highest Ni dissolution rate in both cases, while the process was dependent on pH and ionic strength and is enhanced by complex-forming ligands (Hseu et al., 2017).

Bioleaching is a prospective substitute for conventional metallurgical procedures in the recovery of metals from solid matrices due to its more acceptable environmental profile, efficacy, and metal selectivity. It is a microbial-based method that use living organisms for transforming metal fractions into soluble and ectractable forms. Compared to traditional chemical leaching techniques, bioleaching is more cost-effective, ease for *in situ* leaching, eliminates the need for use of hazardous chemicals, and reduce production of secondary wastes. It can be applied to the range of solid matrices, including mining and electronic waste, industrial sludge, and contaminated soil, making the process more sustainable and circulatory (Rendón-Castrillón et al., 2023). Particular attention has been given to waste materials or soils in which concentration of metals is too low for economical processing by conventional methods. Bioleaching process is impacted by pH, temperature, oxygen supply, and presence of nutrients in the medium. Bacteria *Bacillus* and *Pseudomonas* and fungi *Aspergillus* and *Penicillium* are among the most often used genera for bioleaching purposes (Castro et al., 2000). *Acidithiobacillus thiooxidans*, a native sulfur-oxidizing bacteria applied for bioleaching of metals from contaminated tannery soil in Ranipet, India, at varying starting pH values produced from 59% to 98% of solubilization rate of chromium, zinc, copper, lead, and cadmium (Naresh Kumar and Nagendran, 2009). After the bioleaching process, cadmium, copper, and zinc were entirely eliminated from all mobile soil fractions, remaining only in the residual fraction. These results confirmed that bioleaching can be efficiently applied for the removal of metals from contaminated soils. Investigation of bioleaching nickel from olivine, Ni-bearing mineral present in ultramafic rocks, has been conducted using bacterial and fungal species. Fungi *Aspergillus niger* showed higher nickel leaching quantity in comparison to bacteria *Paenibacillus mucilaginosus* (7.8% in comparison to 5.3%, respectively, in 7-day treatment) (Chiang et al., 2014). Ultrasonic-assisted bioleaching positively influenced fungal growth and showed overall increase of the bioleaching rate for over 50%, showing that olivine can potentially become an economically viable nickel supply with the use of assisted bioleaching.

Electrokinetic remediation is another remediation system able to separate contaminants from the soil. The type and concentration of pollutants, their ionic mobility and solubility, and the physical and chemical characteristics of contaminated soil affect the effects of this process. Electrochemical treatment of washing solution generated after ethylendiaminetetraacetic acid (EDTA) chelate leaching of Cd, Pb, and Zn from contaminated soil was investigated by Finzgar and Lestan (2008) with the aim of acessing the feasibility of two-phase remediation technology for contaminated soils. During EDTA-remediation process 35% of total Cd, 44% of total Pb, and 14% of total Zn content were removed from the soil. Application of electrochemical advanced oxidation process resulted in removal of 90% of Cd, 99% of Pb and 92% of Zn from washing solution, where during the oxidative breakdown of EDTA complexes metals precipitated from the solution or have been deposited on the cathode. The applied method enabled remediation with smaller water requirements and produced safely disposable discharge solution with very low concentration of metals, while deposited metals were available for further recovery.

Specific type of phytoextraction that utilizes hyperaccumulating plants, called phytomining, represents promising technology for metal extraction from metal-rich soils through their uptake and concentration in plant above-ground biomass. After harvesting, biomass of these plants is subjected to various metal extraction technologies, often relying on developing pyrometallurgical or hydrometallurgical procedures (Jovanović et al., 2022). Majority of hyperaccumulating plants are known for ability to accumulate Ni, therefore various aspects of nickel phytomining were particularly investigated in recent decades (Nkrumah et al., 2016). Commercial application of phytomining is in its early stages. Use of Ni hyperaccumulator *Odontarrhena muralis* for phytomining of ultramafic sites in Albania by the Econick spin-off company from the University of Lorraine, France, is one of such examples. By processing nickel recovered from biomass to the various Ni products, financial feasibility of phytomining process can be significantly improved (Barbaroux et al., 2012).

Metal recovery from naturaly or anthropogenically enriched soils is developing field assisting in satisfying the growing needs for utilizing valuable metal resources. This process can reduce the need for traditional mining, provide an opportunity to remediate contaminated soils and improve soil quality, and offer recovery of metal resources from wastes generated after various industrial activities. Certain challenges must be overcome in order to implement the technologies for metal recovery more efficiently: role of soil characteristics in optimization of existing or developing technologies needs to be well recognized, addition of chemicals have to be minimized or they have to be applied in sustainable and circular economy manner, and technologies for recovery of metals need to be low-cost and practical for large-scale applications. Furthermore, it is vital to consider the total environmental impact of these technologies and ensure that they do not cause harm to ecosystems. Collaboration between researchers, industry professionals, and policymakers is essential to address these challenges and promote sustainable metal recovery practices.

2.7 Conclusion

Increased presence of metals in soils, both of natural and anthropogenic origin, resulted in the occurrence of metalliferous soils worldwide. Due to the increasing industrial, mining, agricultural, and urban activities, global metal load and surface under metal-rich soils is constantly increasing. Such soils become source of mobile metals for groundwaters and living organisms. By penetrating into biota, metals from soils influence food chains, crop production, food security, and human health. This demands development and application of new and integrated soil metal remediation technologies, with special focus on those that are environmentally-friendly and cost-effective. Although physical and chemical soil remediation methods can be fast and effective, the costs of their application are usually high and sometimes secondary pollution may occur. Use of plants able to tolerate and accumulate metals from soils (metallophytae and hyperaccumulator plants) paves the way toward wider development of phytoremediation and phytomining approaches for sustainable remediation of metal-contaminated soils and successive increase of soil quality and health. Moreover, coupling with other physical or chemical remediation techniques can help mitigate disadvantages of single remediation technology. Besides presenting spreading environmental burden that requires continual monitoring and sanitation, metalliferous soils are today also recognized as a potential resource for gaining valuable metals. This is especially the case when coupled with management strategies and remediation practices that lead to the synergy of soil quality enhancement and sustainable metal recovery. Invention of environmental-friendly, bio-based, low-cost, and time-efficient strategies for the extraction of various metals from different soil matrices are subjects of ongoing researches and continual efforts in biotechnology development.

References

Abreu, M.M., Matias, M.J., Magalhães, M.C.F., Basto, M.J., 2008. Impacts on water, soil and plants from the abandoned Miguel Vacas copper mine, Portugal. Journal of Geochemical Exploration 96 (2-3), 161–170. Available at 10.1016/j.gexplo.2007.04.012.

Adewumi, A.J., Ogundele, O.D., Villaseñor, M.G., 2024. Hidden hazards in urban soils: a meta-analysis review of global heavy metal contamination (2010–2022), sources and its ecological and health consequences. Sustainable Environment 10 (1), 2293239. Available at https://doi.org/10.1080/27658511.2023.2293239.

Albanese, S., Sadeghi, M., Lima, A., Cicchella, D., Dinelli, E., Valera, P., et al., 2015. GEMAS: Cobalt, Cr, Cu and Ni distribution in agricultural and grazing land soil of

Europe. Journal of Geochemical Exploration 154, 81–93. Available at https://doi.org/10.1016/j.gexplo.2015.01.004.

Andráš, P., Midula, P.l, Matos, J.X., Buccheri, G., Drimal, M., Dirner, V., 2021. Comparison of soil contamination at the selected European Copper Mines. Carpathian Journal of Earth and Environmental Sciences 16 (1), 163–174. Available at 10.26471/cjees/2021/016/164.

Alencar Meira da Silva, H., Davée Guimarães, J.R., 2024. Mercury cyanide complexes and their relevance as environmental contaminants. Chemosphere 350, 141054. Available at https://doi.org/10.1016/j.chemosphere.2023.141054.

Allen, R.L., Weihed, P., 2002. Global comparisons of volcanic-associated massive sulphide districts. Special Publications 204 (1), 13–37. Available at https://doi.org/10.1144/gsl.sp.2002.204.

Álvarez-López, V., Prieto-Fernández, Á., Cabello-Conejo, M.I., Kidd, P.S., 2016. Organic amendments for improving biomass production and metal yield of Ni-hyperaccumulating plants. Science of the Total Environment 370–379. Available at https://doi.org/10.1016/j.scitotenv.2015.12.147.

Asare, M.O., Afriyie, J.O., 2021. Ancient mining and metallurgy as the origin of Cu, Ag, Pb, Hg, and Zn contamination in soils: a review. Water, Air, and Soil Pollution 232 (6), 240. Available at https://doi.org/10.1007/s11270-021-05166-4, http://www.kluweronline.com/issn/0049–6979/.

Barbaroux, R., Plasari, E., Mercier, G., Simonnot, M.O., Morel, J.L., Blais, J.F., 2012. A new process for nickel ammonium disulfate production from ash of the hyperaccumulating plant Alyssum murale. Science of the Total Environment 423, 111–119. Available at https://doi.org/10.1016/j.scitotenv.2012.01.063.

Barbafieri, M., Dadea, C., Tassi, E., Bretzel, F., Fanfani, L., 2011. Uptake of Heavy Metals by Native Species Growing in a Mining Area in Sardinia, Italy: Discovering Native Flora for Phytoremediation. International Journal of Phytoremediation 13 (10), 985–997. Available at 10.1080/15226514.2010.549858.

Baker AJM, Ernst WHO, van der Ent A, Malaisse F, Ginocchio R, Metallophytes: the unique biological resource, its ecology and conservation status in Europe, Central Africa, and Latin America. In: Batty LC, Hallberg KB (eds) Ecology of industrial pollution. Cambridge University Press, New York, 2010, pp 7–40.

Berger, V.I., Singer, D.A., Bliss, J.D., Moring, B.C., 2011. Ni-Co laterite deposits of the world – Database and Gradde and Tonnage Models. Open-File Report 2011-1058, US Geological Survey.

Bini, C., Sartori, G., Wahsha, M., Fontana, S., 2011. Background levels of trace elements and soil geochemistry at regional level in NE Italy. Journal of Geochemical Exploration 109 (1–3), 125–133. Available at https://doi.org/10.1016/j.gexplo.2010.07.008.

Boni, M., Mondillo, N., 2015. The "calamines" and the "others": The great family of supergene nonsulfide zinc ores. Ore Geology Reviews 67, 208–233. Available at https://doi.org/10.1016/j.oregeorev.2014.10.025.

Bonnet, M., Caner, L., Siitari-Kauppi, M., Mazurier, A., Mexias, A.S., Dani, N., et al., 2022. Weathering of Viamão granodiorite, South Brazil. Part 1: Clay minerals formation and increase in total porosity. Geoderma 424, 115968. Available at https://doi.org/10.1016/j.geoderma.2022.115968.

Borg, G., Karner, K., Buxton, M., Armstrong, R., v.d. Merwe, S.W., 2003. Geology of the Skorpion Supergene Zinc Deposit, Southern Namibia. Economic Geology 98 (4), 749–771. Available at https://doi.org/10.2113/gsecongeo.98.4.749.

Borst, A.M., Smith, M.P., Finch, A.A., Estrade, G., Villanova-de-Benavent, C., Nason, P., et al., 2020. Adsorption of rare earth elements in regolith-hosted clay deposits. Nature Communications 11 (1), 4386. Available at https://doi.org/10.1038/s41467-020-17801-5, http://www.nature.com/ncomms/index.html.

Brooks, R.R., Reeves, R.D., Morrison, R.S., Malaisse, F., 1980. Hyperaccumulation of copper and cobalt: a review. Bulletin de la Société Royale de Botanique de Belgique 13, 166–172. Available at https://doi.org/10.2307/20793840.

Bustamante-Rúa, M.O., Gonzalez-Arias, S.D., Bustamante-Baena, P., 2020. Nickel laterite concentration through a non-conventional method with surface sulfidization. DYNA (Colombia) 87 (215), 18–27. Available at https://doi.org/10.15446/dyna.v87n215.85981, http://www.scielo.org.co/pdf/dyna/v87n215/2346-2183-dyna-87–215-18.pdf.

Cachada, A., Rocha-Santos, T., Duarte, A.C., 2018. Soil and pollution: an introduction to the main issues. Soil Pollution From Monitoring to Remediation. Academic Press, pp. 1–28.

Cao, X., Gao, X., Zeng, X., Ma, Y., Gao, Y., Baeyens, W., et al., 2021. Seeking for an optimal strategy to avoid arsenic and cadmium over-accumulation in crops: soil management vs cultivar selection in a case study with maize. Chemosphere 272, 129891. Available at https://doi.org/10.1016/j.chemosphere.2021.129891.

Cao, J., Xie, C.-yu, Hou, Z.-ru, 2022. Spatiotemporal distribution patterns and risk characteristics of heavy metal pollutants in the soil of lead–zinc mines. Environmental Sciences Europe 34 (1). Available at 10.1186/s12302-022-00607-1.

Carmo, F.F.D., Kamino, L.H.Y., Junior, R.T., Campos, I.C.D., Carmo, F.F.D., Silvino, G., et al., 2017. Fundão tailings dam failures: the environment tragedy of the largest technological disaster of Brazilian mining in global context. Perspectives in Ecology and Conservation 15 (3), 145–151. Available at https://doi.org/10.1016/j.pecon.2017.06.002, http://www.elsevier.com/locate/issn/25300644.

Castro, I.M., Fietto, J.L.R., Vieira, R.X., Trópia, M.J.M., Campos, L.M.M., Paniago, E.B., et al., 2000. Bioleaching of zinc and nickel from silicates using Aspergillus niger cultures. Hydrometallurgy 57 (1), 39–49. Available at https://doi.org/10.1016/s0304–386x(00)00088-8.

Chen, X., Duan, D., Zhang, Y., Zhou, F., Yuan, X., Wu, Y., 2023. Genesis of the giant Huoshaoyun non-sulfide zinc–lead deposit in Karakoram, Xinjiang: constraints from mineralogy and trace element geochemistry. Minerals 13 (7), 842. Available at https://doi.org/10.3390/min13070842.

Chen, P., Ilton, E.S., Wang, Z., Rosso, K.M., Zhang, X., 2024. Global rare earth element resources: a concise review. Applied Geochemistry 175, 106158. Available at https://doi.org/10.1016/j.apgeochem.2024.106158.

Chen, L., Wang, J., Beiyuan, J., Guo, X., Wu, H., Fang, L., 2022. Environmental and health risk assessment of potentially toxic trace elements in soils near uranium (U) mines: a global meta-analysis. Science of the Total Environment 816, 151556. Available at https://doi.org/10.1016/j.scitotenv.2021.151556.

Chen, L., Zhou, M., Wang, J., Zhang, Z., Duan, C., Wang, X., et al., 2022. A global meta-analysis of heavy metal(loid)s pollution in soils near copper mines: evaluation of pollution level and probabilistic health risks. Science of the Total Environment 835, 155441. Available at https://doi.org/10.1016/j.scitotenv.2022.155441.

Chiang, Y.W., Santos, R.M., Audenaerde, A.V., Monballiu, A., Van Gerven, T., Meesschaert, B., 2014. Chemoorganotrophic bioleaching of olivine for nickel recovery. Minerals 4 (2), 553–564. Available at https://doi.org/10.3390/min4020553, http://www.mdpi.com/2075-163X/4/2/541/pdf.

Christou, A., Theologides, C., Costa, C., Kalavrouziotis, I.K., Varnavas, S.P., 2017. Assessment of toxic heavy metals concentrations in soils and wild and cultivated plant species in Limni abandoned copper mining site, Cyprus. Journal of Geochemical Exploration 178, 16–22. Available at 10.1016/j.gexplo.2017.03.012.

Cooper, J., Sanderson, R., Cakmak, I., Ozturk, L., Shotton, P., Carmichael, A., et al., 2011. Effect of organic and conventional crop rotation, fertilization, and crop protection practices on metal contents in wheat (triticum aestivum). Journal of Agricultural and Food Chemistry 59 (9), 4715–4724. Available at https://doi.org/10.1021/jf104389m.

Coppola, V., Boni, M., Gilg, H.A., Balassone, G., Dejonghe, L., 2008. The "calamine" nonsulfide Zn-Pb deposits of Belgium: petrographical, mineralogical and geochemical characterization. Ore Geology Reviews 33 (2), 187–210. Available at https://doi.org/10.1016/j.oregeorev.2006.03.005.

Coppola, V., Boni, M., Gilg, A.H., Strzelska-Smakowska, B., 2009. Nonsulfide zinc deposits in the Silesia-Cracow district, Southern Poland. Mineralium Deposita 44 (5), 559–580. Available at https://doi.org/10.1007/s00126-008-0220-4.

Couto, F.R., Ferreira, A.D.M., Pontes, P.P., Marques, A.R., 2021. Physical, chemical and microbiological characterization of the soils contaminated by iron ore tailing mud after Fundão Dam disaster in Brazil. Applied Soil Ecology 158, 103811. Available at https://doi.org/10.1016/j.apsoil.2020.103811, https://www.elsevier.com/inca/publications/store/5/2/4/5/1/8/index.htt.

D'Amico, M.E., Calabrese, F., Rossetti, A., Previtali, F., 2009. Heavy metals and biological properties of subalpine soils on ophiolites in the Italian Western Alps. Northeastern Naturalist 16, 193–214. Available at https://doi.org/10.1656/045.016.0516.

del Real, I., Reich, M., Simon, A.C., Deditius, A., Barra, F., Rodríguez-Mustafa, M.A., et al., 2023. Formation of giant iron oxide-copper-gold deposits by superimposed episodic hydrothermal pulses. Scientific Reports 13 (1), 12041. Available at https://doi.org/10.1038/s41598-023-37713-w, https://www.nature.com/srep/.

Elias, M., 2002. Nickel Laterite Deposits-Geological Overview, Resources and Exploitation. Centre for Ore Deposit Research, University of Tasmania, Special Publication, No. 4.

Eliopoulos, D.G., Economou-Eliopoulos, M., Apostolikas, A., Golightly, J.P., 2012. Geochemical features of nickel-laterite deposits from the Balkan Peninsula and Gordes, Turkey: the genetic and environmental significance of arsenic. Ore Geology Reviews 48, 413–427. Available at https://doi.org/10.1016/j.oregeorev.2012.05.008.

Finzgar, N., Lestan, D., 2008. The two-phase leaching of Pb, Zn and Cd contaminated soil using EDTA and electrochemical treatment of the washing solution. Chemosphere 73 (9), 1484–1491. Available at https://doi.org/10.1016/j.chemosphere.2008.07.043.

Fernandez-Turiel, J.L., Aceñolaza, P., Medina, M.E., Llorens, J.F., Sardi, F., 2001. Assessment of a Smelter Impact Area Using Surface Soils and Plants. Environmental Geochemistry and Health 23 (1), 65–78. Available at 10.1023/a:1011071704610.

Galey, M.L., van der Ent, A., Iqbal, M.C.M., Rajakaruna, N., 2017. Ultramafic geoecology of South and Southeast Asia. Botanical Studies 58 (1), 18. Available at https://doi.org/10.1186/s40529-017-0167-9.

Ghaderian, S.M., Ghotbi Ravandi, A.A., 2012. Accumulation of copper and other heavy metals by plants growing on Sarcheshmeh copper mining area, Iran. Journal of Geochemical Exploration 123, 25–32. Available at 10.1016/j.gexplo.2012.06.022.

Galán, E., González, I., Fernández-Caliani, J.C., 2002. Residual pollution load of soils impacted by the Aznalcóllar (Spain) mining spill after clean-up operations. Science of the Total Environment 286 (1–3), 167–179. Available at https://doi.org/10.1016/s0048–9697(01)00974-3.

Gibson, N., Meissner, R., Markey, A.S., Thompson, W.A., 2012. Patterns of plant diversity in ironstone ranges in arid south western Australia. Journal of Arid Environments 77 (1), 25–31. Available at https://doi.org/10.1016/j.jaridenv.2011.08.021.

Goldenberg, M., Reddy, K.R., 2014. Sustainability assessment of excavation and disposal versus in situ stabilization of heavy metal-contaminated soil at a superfund site in Illinois. Geo-Congress 2014 Technical Papers: Geo-Characterization and Modeling for

Sustainability GSP 234, pp. 2245–2254http://ascelibrary.org/ doi:10.1061/9780784413272.219.

Gole, M.J., Barnes, S.J., 2020. The association between Ni-Cu-PGE sulfide and Ni-Co lateritic ores and volcanic facies within the komatiites of the 2.7 Ga East Yilgarn Craton Large Igneous Province, Western Australia. Ore Geology Reviews 116, 103231. Available at https://doi.org/10.1016/j.oregeorev.2019.103231, http://www.sciencedirect.com/science/journal/01691368.

Golia, E.E., 2023. The impact of heavy metal contamination on soil quality and plant nutrition. Sustainable management of moderate contaminated agricultural and urban soils, using low cost materials and promoting circular economy. Sustainable Chemistry and Pharmacy 33, 101046. Available at https://doi.org/10.1016/j.scp.2023.101046.

Gonnelli, C., Renella, G., 2012. Chromium and nickel. In: Alloway, B. (Ed.), Heavy Metals in Soils: Trace Metals and Metalloids in Soils and their Bioavailability. Springer, Dordrecht-Heidelberg-New York-London, pp. 313–333.

Graham, R.C., Rossi, A.M., Hubbert, K.R., 2010. Rock to regolith conversion: producing hospitable substrates for terrestrial ecosystems. GSA Today a Publication of the Geological Society of America 20 (2), 4–9. Available at https://doi.org/10.1130/GSAT57A.1, http://www.geosociety.org/gsatoday/archive/20/2/pdf/i1052–5173-20-2-4.pdf.

Grosche, A., Klemd, R., Denkel, K., Keith, M., Haase, K.M., Voudouris, P.C., et al., 2023. Sources, transport, and deposition of metal(loid)s recorded by sulfide and rock geochemistry: constraints from a vertical profile through the epithermal Profitis Ilias Au prospect, Milos Island, Greece. Mineralium Deposita 58 (6), 1101–1122. Available at https://doi.org/10.1007/s00126-023-01170-2, https://www.springer.com/journal/126.

Groves, D.I., Goldfarb, R.J., Robert, F., Hart, C.J.R., 2003. Gold deposits in metamorphic belts: overview of current understanding,outstanding problems, future research, and exploration significance. Economic Geology 98 (1), 1–29. Available at https://doi.org/10.2113/gsecongeo.98.1.1.

Guarino, C., Sciarrillo, R., 2017. Effectiveness of in situ application of an Integrated Phytoremediation System (IPS) by adding a selected blend of rhizosphere microbes to heavily multi-contaminated soils. Ecological Engineering 99, 70–82. Available at https://doi.org/10.1016/j.ecoleng.2016.11.051.

Guo, Z., Megharaj, M., Beer, M., Ming, H., Mahmudur Rahman, M., Wu, W., et al., 2009. Heavy metal impact on bacterial biomass based on DNA analyses and uptake by wild plants in the abandoned copper mine soils. Bioresource Technology 100 (17), 3831–3836. Available at 10.1016/j.biortech.2009.02.043.

Hagemann, S.G., Angerer, T., Duuring, P., Rosière, C.A., Figueiredo e Silva, R.C., Lobato, L., et al., 2016. BIF-hosted iron mineral system: a review. Ore Geology Reviews 76, 317–359. Available at https://doi.org/10.1016/j.oregeorev.2015.11.004.

Han, L., Zhai, Y., Chen, R., Fan, Y., Liu, Z., Zhao, Y., et al., 2023. Characteristics of soil arsenic contamination and the potential of pioneer plants for arsenic remediation in gold mine tailings. Toxics 11 (12), 1025. Available at https://doi.org/10.3390/toxics11121025.

Hou, D., Jia, X., Wang, L., McGrath, S.P., Zhu, Y.-G., Hu, Q., et al., 2025. Global soil pollution by toxic metals threatens agriculture and human health. Science (New York, N.Y.) 388 (6744), 316–321. Available at https://doi.org/10.1126/science.adr5214.

Hou, X., Parent, M., Savard, M.M., Tassé, N., Bégin, C., Marion, J., 2006. Lead concentrations and isotope ratios in the exchangeable fraction: tracing soil contamination near a copper smelter. Geochemistry: Exploration, Environment, Analysis 6 (2–3), 229–236. Available at https://doi.org/10.1144/1467–7873/05–092, http://geea.lyellcollection.org/.

Hou, Y., Zhao, Y., Lu, J., Wei, Q., Zang, L., Zhao, X., 2023. Environmental contamination and health risk assessment of potentially toxic trace metal elements in soils near gold mines: a global meta-analysis. Environmental Pollution 330, 121803. Available at https://doi.org/10.1016/j.envpol.2023.121803.

Hseu, Z.Y., Su, Y.C., Zehetner, F., Hsi, H.C., 2017. Leaching potential of geogenic nickel in serpentine soils from Taiwan and Austria. Journal of Environmental Management 186, 151–157. Available at https://doi.org/10.1016/j.jenvman.2016.02.034, https://www.sciencedirect.com/science/journal/03014797.

Huang, Y., Wang, L., Wang, W., Li, T., He, Z., Yang, X., 2019. Current status of agricultural soil pollution by heavy metals in China: a meta-analysis. Science of the Total Environment 651 (Pt 2), 3034–3042. Available at https://doi.org/10.1016/j.scitotenv.2018.10.185.

Huston, D.L., Stevens, B., Southgate, P.N., Muhling, P., Wyborn, L., 2006. Australian Zn–Pb–Ag ore-forming systems: a review and analysis. Economic Geology 101 (6), 1117–1157. Available at https://doi.org/10.2113/gsecongeo.101.6.1117.

Jakovljevic, K., Bani, A., Pavlova, D., Konstantinou, M., Dimitrakopoulos, P., Kyrkas, D., et al., 2022. Hyperaccumulator plant discoveries in the Balkans: accumulation, distribution, and practical applications. Botanica Serbica 46 (2), 161–178. Available at https://doi.org/10.2298/botserb2202161j.

Jakovljević, K., Mišljenović, T., Bačeva Andonovska, K., Echevarria, G., Baker, A.J.M., Brueckner, D., et al., 2023. Thallium hyperaccumulation status of the violets of the Allchar arsenic-thallium deposit (North Macedonia) confirmed through synchrotron μXRF imaging. Metallomics 15 (11), mfad063. Available at https://doi.org/10.1093/mtomcs/mfad063, https://academic.oup.com/metallomics/issue/13/1.

Jakovljević, K., Mišljenović, T., Bačeva Andonovska, K., Echevarria, G., Charrois, L., van der Ent, Antony, 2023. Living at the edge of life: metallophytes from the most toxic arsenic-thallium tailings in the world (Allchar, North Macedonia). Plant and Soil 497 (1-2), 413–428. Available at 10.1007/s11104-023-06404-w.

Jakovljević, K., Mišljenović, T., Savović, J., Ranković, D., Ranđelović, D., Mihailović, N., et al., 2020. Accumulation of trace elements in *Tussilago farfara* colonizing post-flotation tailing sites in Serbia. Environmental Science and Pollution Research 27 (4), 4089–4103. Available at https://doi.org/10.1007/s11356-019-07010-z.

Jiang, S., Duan, L., Dai, G., Shu, Y., 2021. Immobilization of heavy metal(loid)s in acid paddy soil by soil replacement-biochar amendment technology under normal wet condition. Environmental Science and Pollution Research 28 (48), 68886–68896. Available at https://doi.org/10.1007/s11356-021-14757-x.

Jovanović, G., Ranđelović, D., Marković, B., Sokić, M., 2022. Overview of technologies for Zn extraction from hyperaccumulating plants: current state of research and future directions. Journal of Mining and Metallurgy A: Mining 58 (1), 29–38. Available at https://doi.org/10.5937/jmma2201029j.

Juilleret, J., Dondeyne, S., Vancampenhout, K., Deckers, J., Hissler, C., 2016. Mind the gap: a classification system for integrating the subsolum into soil surveys. Geoderma 264, 332–339. Available at https://doi.org/10.1016/j.geoderma.2015.08.031.

Kabir, E., Ray, S., Kim, K.H., Yoon, H.O., Jeon, E.C., Kim, Y.S., et al., 2012. Current status of trace metal pollution in soils affected by industrial activities. The Scientific World Journal 2012, 916705. Available at https://doi.org/10.1100/2012/916705.

Kabata-Pendias, A, Pendias, H. Trace Elements in Soils and Plants. 4th Edition, 2011, CRC Press, Boca Raton.

Kama, R., Li, S., Nabi, F., Aidara, M., Huang, P., Li, Z., et al., 2024. Hyperaccumulators' diversity enhances Cd-contaminated soil restoration and reduces rice Cd uptake under an intercropping system. ACS Omega 9 (26), 28784–28790. Available at https://doi.org/10.1021/acsomega.4c03107.

Karbowska, B., 2016. Presence of thallium in the environment: sources of contaminations, distribution and monitoring methods. Environmental Monitoring and Assessment 188 (11), 640. Available at https://doi.org/10.1007/s10661-016-5647-y.

Kelepertzis, E., 2014. Accumulation of heavy metals in agricultural soils of Mediterranean: Insights from Argolida basin, Peloponnese, Greece. Geoderma 221-222, 82–90. Available at 10.1016/j.geoderma.2014.01.007.

Kierczak, J., Pietranik, A., Pędziwiatr, A., 2021. Ultramafic geoecosystems as a natural source of Ni, Cr, and Co to the environment: a review. Science of the Total Environment 755 (Pt 1), 142620. Available at https://doi.org/10.1016/j.scitotenv.2020.142620.

Larocque, A.C.L., Rasmussen, P.E., 1998. An overview of trace metals in the environment, from mobilization to remediation. Environmental Geology 33 (2–3), 85–91. Available at https://doi.org/10.1007/s002540050227.

Lago-Vila, M., Arenas-Lago, D., Rodríguez-Seijo, A., Andrade Couce, M.L., Vega, F.A., 2015. Cobalt, chromium and nickel contents in soils and plants from a serpentinite quarry. Solid Earth 6 (1), 323–335. Available at 10.5194/se-6-323-2015.

Lewchuk, M.T., Leach, D.L., Kelley, K.D., A. Symons, D.T., 2004. Paleomagnetism of the Red Dog Zn-Pb massive sulfide deposit in northern Alaska. Economic Geology 99 (7), 1555–1567. Available at https://doi.org/10.2113/gsecongeo.99.7.1555.

Lin, Shui-Quan, Wang, Guang-Jin, Liu, Wen-Lian, Zhao, Bing, Shen, Ying-Ming, Wang, Meng-Lai, et al., 2022. Regional Distribution and Causes of Global Mine Tailings Dam Failures. Metals 12 (6), 905. Available at 10.3390/met12060905.

Luo, C., Liu, C., Wang, Y., Liu, X., Li, F., Zhang, G., et al., 2011. Heavy metal contamination in soils and vegetables near an e-waste processing site, south China. Journal of Hazardous Materials 186 (1), 481–490. Available at 10.1016/j.jhazmat.2010.11.024.

Lin, S.Q., Wang, G.J., Liu, W.L., Zhao, B., Shen, Y.M., Wang, M.L., et al., 2022. Regional distribution and causes of global mine tailings dam failures. Metals 12 (6), 905. Available at https://doi.org/10.3390/met12060905, https://www.mdpi.com/2075–4701/12/6/905/pdf?version=1653533407.

Liu, S.L., Fan, H.R., Liu, X., Meng, J., Butcher, A.R., Yann, L., et al., 2023. Global rare earth elements projects: new developments and supply chains. Ore Geology Reviews 157, 105428. Available at https://doi.org/10.1016/j.oregeorev.2023.105428, http://www.sciencedirect.com/science/journal/01691368.

Liu, H., Liao, R., Zhang, L., Li, C., Sun, W., 2020. Plate subduction, oxygen fugacity, and mineralization. Journal of Oceanology and Limnology 38 (1), 64–74. Available at https://doi.org/10.1007/s00343-019-8339-y.

Liu, J., Luo, X., Sun, Y., Tsang, D.C.W., Qi, J., Zhang, W., et al., 2019. Thallium pollution in China and removal technologies for waters: a review. Environment International 126, 771–790. Available at https://doi.org/10.1016/j.envint.2019.01.076, https://www.elsevier.com/locate/envint.

Liu, N., Zhao, J., Du, J., Hou, C., Zhou, X., Chen, J., et al., 2024. Non-phytoremediation and phytoremediation technologies of integrated remediation for water and soil heavy metal pollution: a comprehensive review. Science of the Total Environment 948, 174237. Available at https://doi.org/10.1016/j.scitotenv.2024.174237.

Lwin, C.S., Seo, B.H., Kim, H.U., Owens, G., Kim, K.R., 2018. Application of soil amendments to contaminated soils for heavy metal immobilization and improved soil quality—a critical review. Soil Science and Plant Nutrition 64 (2), 156–167. Available at https://doi.org/10.1080/00380768.2018.1440938, http://www.tandfonline.com/toc/tssp20/current.

Mansoor, S., Kour, N., Manhas, S., Zahid, S., Wani, O.A., Sharma, V., et al., 2021. Biochar as a tool for effective management of drought and heavy metal toxicity.

Chemosphere 271, 129458. Available at https://doi.org/10.1016/j.chemosphere.2020.129458, https://www.elsevier.com/locate/chemosphere.
Mahdavian, K., Ghaderian, S.M., Torkzadeh-Mahani, M., 2015. Accumulation and phytoremediation of Pb, Zn, and Ag by plants growing on Koshk lead–zinc mining area, Iran. Journal of Soils and Sediments 17 (5), 1310–1320. Available at 10.1007/s11368-015-1260-x.
Mao, X., Han, F.X., Shao, X., Guo, K., McComb, J., Arslan, Z., et al., 2016. Electrokinetic remediation coupled with phytoremediation to remove lead, arsenic and cesium from contaminated paddy soil. Ecotoxicology and Environmental Safety 125, 16–24. Available at https://doi.org/10.1016/j.ecoenv.2015.11.021, https://www.elsevier.com/inca/publications/store/6/2/2/8/1/9/index.htt.
Misra, Kula C., 2000. Skarn Deposits. Understanding Mineral Deposits. Springer Netherlands, pp. 414–449. Available at 10.1007/978-94-011-3925-0_9.
Marescotti, P., Comodi, P., Crispini, L., Gigli, L., Zucchini, A., Fornasaro, S., 2019. Potentially toxic elements in ultramafic soils: a study from metamorphic ophiolites of the Voltri Massif (Western Alps, Italy). Minerals 9 (8), 502. Available at https://doi.org/10.3390/min9080502.
Marrugo-Negrete, J., Pinedo-Hernández, J., Díez, S., 2017. Assessment of heavy metal pollution, spatial distribution and origin in agricultural soils along the Sinú River Basin, Colombia. Environmental Research 154, 380–388. Available at 10.1016/j.envres.2017.01.021.
Matanzas, N., Afif, E., Díaz, T.E., Gallego, J.L.R., 2021. Screening of pioneer metallophyte plant species with phytoremediation potential at a severely contaminated Hg and As mining site. Environments 8 (7), 63. Available at https://doi.org/10.3390/environments8070063.
Maus, V., Giljum, S., Gutschlhofer, J., da Silva, D.M., Probst, M., Gass, S.L.B., et al., 2020. A global-scale data set of mining areas. Scientific Data 7 (1), 289. Available at https://doi.org/10.1038/s41597-020-00624-w, https://www.nature.com/sdata/.
Ferraz, M.C.M.A., Lourençlo, J.C.N., 2000. The influence of organic matter content of contaminated soils on the leaching rate of heavy metals. Environmental Progress 19 (1), 53–58. Available at https://doi.org/10.1002/ep.670190118.
McMartin, I., Henderson, P.J., Nielsen, E., 1999. Impact of a base metal smelter on the geochemistry of soils of the Flin Flon region, Manitoba and Saskatchewan. Canadian Journal of Earth Sciences 36 (2), 141–160. Available at https://doi.org/10.1139/e98-001.
Mielke, J.E., 1979. Composition of the Earth's crust and distribution of elements. Review on Research and Modern Problems in Geochemistry 16. UNESCO Reportt, pp. 9–37.
Mishra, B., Bernhardt, H.J., 2009. Metamorphism, graphite crystallinity, and sulfide anatexis of the Rampura-Agucha massive sulfide deposit, northwestern India. Mineralium Deposita 44 (2), 183–204. Available at https://doi.org/10.1007/s00126-008-0208-0.
Misra, K.C., 2000. Skarn Deposits. Understanding Mineral Deposits. Springer, Dordrecht, pp. 414–449. Available at https://doi.org/10.1007/978-94-011–3925-0_9.
Monteiro, L.V.S., Bettencourt, J.S., Juliani, C., de Oliveira, T.F., 2007. Nonsulfide and sulfide-rich zinc mineralizations in the Vazante, Ambrósia and Fagundes deposits, Minas Gerais, Brazil: mass balance and stable isotope characteristics of the hydrothermal alterations. Gondwana Research 11 (3), 362–381. Available at https://doi.org/10.1016/j.gr.2006.04.017, http://www.sciencedirect.com/science/journal/1342937X.
Monterroso, C., Rodríguez, F., Chaves, R., Diez, J., Becerra-Castro, C., Kidd, P.S., et al., 2014. Heavy metal distribution in mine-soils and plants growing in a Pb/Zn-mining area in NW Spain. Applied Geochemistry 44, 3–11. Available at 10.1016/j.apgeochem.2013.09.001.

Mudd, G.M., 2010. Global trends and environmental issues in nickel mining: sulfides versus laterites. Ore Geology Reviews 38 (1–2), 9–26. Available at https://doi.org/10.1016/j.oregeorev.2010.05.003.

Mukube, P., Hitzman, M., Machogo-Phao, L., Syampungani, S., 2024. Geochemistry of terrestrial plants in the Central African Copperbelt: implications for sediment hosted copper-cobalt exploration. Minerals 14 (3), 294. Available at https://doi.org/10.3390/min14030294.

Naresh Kumar, R., Nagendran, R., 2009. Fractionation behavior of heavy metals in soil during bioleaching with *Acidithiobacillus thiooxidans*. Journal of Hazardous Materials 169 (1–3), 1119–1126. Available at https://doi.org/10.1016/j.jhazmat.2009.04.069.

Nde, S.C., Felicite, O.M., Aruwajoye, G.S., Palamuleni, L.G., 2024. A meta-analysis and experimental survey of heavy metals pollution in agricultural soils. Journal of Trace Elements and Minerals 9, 100180. Available at https://doi.org/10.1016/j.jtemin.2024.100180, https://www.sciencedirect.com/journal/journal-of-trace-elements-and-minerals.

Nkrumah, P.N., Baker, A.J.M., Chaney, R.L., Erskine, P.D., Echevarria, G., Morel, J.L., et al., 2016. Current status and challenges in developing nickel phytomining: an agronomic perspective. Plant and Soil 406 (1–2), 55–69. Available at https://doi.org/10.1007/s11104-016-2859-4, http://www.wkap.nl/journalhome.htm/0032–079X.

Ondrasek, G., Shepherd, J., Rathod, S., Dharavath, R., Rashid, M.I., Brtnicky, M., et al., 2025. Metal contamination - a global environmental issue: sources, implications & advances in mitigation. RSC Advances 15 (5), 3904–3927. Available at https://doi.org/10.1039/d4ra04639k, http://pubs.rsc.org/en/journals/journal/ra.

Olko, A., Abratowska, A., Żyłkowska, J., Wierzbicka, M., Tukiendorf, A., 2008. Armeria maritima from a calamine heap—Initial studies on physiologic–metabolic adaptations to metal-enriched soil. Ecotoxicology and Environmental Safety 69 (2), 209–218. Available at 10.1016/j.ecoenv.2007.01.010.

Park, J.W., Campbell, I.H., Chiaradia, M., Hao, H., Lee, C.T., 2021. Crustal magmatic controls on the formation of porphyry copper deposits. Nature Reviews Earth and Environment 2 (8), 542–557. Available at https://doi.org/10.1038/s43017-021-00182-8, nature.com/natreveearthenviron/.

Parviainen, A., Loukola-Ruskeeniemi, K., 2019. Environmental impact of mineralised black shales. Earth-Science Reviews 192, 65–90. Available at https://doi.org/10.1016/j.earscirev.2019.01.017.

Pratas, J., Favas, P. J.C., D'Souza, R., Varun, M., Paul, M.S., 2013. Phytoremedial assessment of flora tolerant to heavy metals in the contaminated soils of an abandoned Pb mine in Central Portugal. Chemosphere 90 (8), 2216–2225. Available at 10.1016/j.chemosphere.2012.09.079.

Pędziwiatr, A., Kierczak, J., Waroszewski, J., Ratié, G., Quantin, C., Ponzevera, E., 2018. Rock-type control of Ni, Cr, and Co phytoavailability in ultramafic soils. Plant and Soil 423 (1–2), 339–362. Available at https://doi.org/10.1007/s11104-017-3523-3.

Pourret, O., Lange, B., Houben, D., Colinet, G., Shutcha, M., Faucon, M.-P., 2015. Modeling of cobalt and copper speciation in metalliferous soils from Katanga (Democratic Republic of Congo). Journal of Geochemical Exploration 149, 87–96. Available at 10.1016/j.gexplo.2014.11.011.

Purwadi, I., Erskine, P.D., Hutahaean, B.P., Wijaya, T.R., Nurtjahya, E., van der Ent, A., 2025. Rare earth elements (REEs) in soils and plants of Bangka Island (Indonesia) focussing on (hyper)accumulation. Plant and Soil 507 (1), 417–431. Available at https://doi.org/10.1007/s11104-024-06735-2, https://www.springer.com/journal/11104.

Rajabi, A., Alfonso, P., Canet, C., Rastad, E., Niroomand, S., Modabberi, S., et al., 2020. The world-class Koushk Zn-Pb deposit, Central Iran: a genetic model for vent-proximal shale-hosted massive sulfide (SHMS) deposits—based on paragenesis and stable isotope geochemistry. Ore Geology Reviews 124, 103654. Available at https://doi.org/10.1016/j.oregeorev.2020.103654.

Ranđelović, D., Gajić, G., Mutić, J., Pavlović, P., Mihailović, N., Jovanović, S., 2016. Ecological potential of Epilobium dodonaei Vill. for restoration of metalliferous mine wastes. Ecological Engineering 95, 800–810. Available at https://doi.org/10.1016/j.ecoleng.2016.07.015.

Ranđelović, D., Jakovljević, K., Šinžar-Sekulić, J., Kuzmič, F., Šilc, U., 2024. Recognising the role of ruderal species in restoration of degraded lands. Science of the Total Environment 938, 173104. Available at https://doi.org/10.1016/j.scitotenv.2024.173104.

Ranđelović, D., Jovanović, S., 2023. Understanding the role of ruderal plant species in restoration of degraded lands,. In: Pandey, V.C. (Ed.), Bio-Inspired Land Remediation. Springer, pp. 31–67 Chapter 2.

Ranđelović, D., Mutić, J., Marjanović, P., Đorđević, T., Kašanin-Grubin, M., 2020. Geochemical distribution of selected elements in flotation tailings and soils/sediments from the dam spill at the abandoned antimony mine Stolice, Serbia. Environmental Science and Pollution Research 27 (6), 6253–6268. Available at https://doi.org/10.1007/s11356-019-07348-4.

Ranđelović, D., Mihailović, N., Jovanović, S., 2019. Potential of Equisetum ramosissimum Desf. for remediation of antimony flotation tailings: a case study. International Journal of Phytoremediation 21 (7), 707–713. Available at https://doi.org/10.1080/15226514.2018.1556590.

Ranđelović, D., Stanković, S., Mihailović, N., Leštan, D., 2015. Remediation of copper from copper mine wastes and contaminated soils using (S,S)-ethylenediaminedisuccinic acid and acidophilic bacteria. Bioremediation Journal 19 (3), 231–238. Available at https://doi.org/10.1080/10889868.2014.995370.

Reeves, R.D., Baker, A.J.M., Borhidi, A., Berazaín, R., 1999. Nickel hyperaccumulation in the serpentine flora of Cuba. Annals of Botany 83 (1), 29–38. Available at https://doi.org/10.1006/anbo.1998.0786, http://aob.oxfordjournals.org/.

Reeves, R.D., Baker, A.J.M., Jaffré, T., Erskine, P.D., Echevarria, G., van der Ent, A., 2018. A global database for plants that hyperaccumulate metal and metalloid trace elements. New Phytologist 218 (2), 407–411. Available at https://doi.org/10.1111/nph.14907, http://onlinelibrary.wiley.com/journal/10.1111/(ISSN)1469-8137.

Rendón-Castrillón, L., Ramírez-Carmona, M., Ocampo-López, C., Gómez-Arroyave, L., 2023. Bioleaching techniques for sustainable recovery of metals from solid matrices. Sustainability 15 (13), 10222. Available at https://doi.org/10.3390/su151310222.

Rodríguez, L., Ruiz, E., Alonso-Azcárate, J., Rincón, J., 2009. Heavy metal distribution and chemical speciation in tailings and soils around a Pb–Zn mine in Spain. Journal of Environmental Management 90 (2), 1106–1116. Available at https://doi.org/10.1016/j.jenvman.2008.04.007.

Rouhani, A., Gutiérrez, M., Newton, R.A., Al Souki, K.S., 2025. An overview of potentially toxic element pollution in soil around lead–zinc mining areas. Environmental Reviews 33, 1–16. Available at https://doi.org/10.1139/er-2024-0052, https://www.nrcresearchpress.com/loi/er.

Singer, D.A., 2017. Future copper resources. Ore Geology Reviews 86, 271–279. Available at https://doi.org/10.1016/j.oregeorev.2017.02.022, http://www.sciencedirect.com/science/journal/01691368.

Soh Tamehe, L., Zhao, Y., Xu, W., Gao, J., 2024. Ni(Co) laterite deposits of southeast asia: a review and perspective. Minerals 14 (2), 134. Available at https://doi.org/10.3390/min14020134.

Sonter, L.J., Dade, M.C., Watson, J.E.M., Valenta, R.K., 2020. Renewable energy production will exacerbate mining threats to biodiversity. Nature Communications 11 (1), 4174. Available at https://doi.org/10.1038/s41467-020-17928-5, http://www.nature.com/ncomms/index.html.

Squire, R.J., Keays, R.R., 2024. The role of supermountain belts and climatic controls on the genesis of copper deposits in the Kupferschiefer and the Central African Copperbelt. Mineralium Deposita 59 (4), 717–732. Available at https://doi.org/10.1007/s00126-023-01227-2, https://www.springer.com/journal/126.

Stevanovic, V., Tan, K., Iatrou, G., 2003. Distribution of the endemic Balkan flora on serpentine I?obligate serpentine endemics. Plant Systematics and Evolution 242 (1–4), 149–170. Available at https://doi.org/10.1007/s00606-003-0044-8.

Sun, W., Huang, R.-fang, Li, H., Hu, Y.-bin, Zhang, C.-chan, Sun, S.-jun, et al., 2015. Porphyry deposits and oxidized magmas. Ore Geology Reviews 65 (1), 97–131. Available at https://doi.org/10.1016/j.oregeorev.2014.09.004.

Swęd, M., Potysz, A., Bartz, W., Siepak, M., 2023. Element dissolution from Zn-bearing rocks treated with chemical and biotic agents: a prospective circular economy strategy for metal recovery from secondary resources. Geochemistry 83 (4), 126008. Available at https://doi.org/10.1016/j.chemer.2023.126008.

Tang, L., Werner, T.T., 2023. Global mining footprint mapped from high-resolution satellite imagery. Communications Earth and Environment 4 (1), 134. Available at https://doi.org/10.1038/s43247-023-00805-6, https://www.nature.com/commsenv/.

Verma, S., Bhatt, P., Verma, A., Mudila, H., Prasher, P., Rene, E.R., 2023. Microbial technologies for heavy metal remediation: effect of process conditions and current practices. Clean Technologies and Environmental Policy 25 (5), 1485–1507. Available at https://doi.org/10.1007/s10098-021-02029-8, https://www.springer.com/journal/10098.

Verma, S., Kuila, A., 2019. Bioremediation of heavy metals by microbial process. Environmental Technology & Innovation 14, 100369. Available at https://doi.org/10.1016/j.eti.2019.100369.

Wang, C., Tan, H., Li, H., Xie, Y., Liu, H., Xu, F., et al., 2020. Mechanism study of Chromium influenced soil remediated by an uptake-detoxification system using hyperaccumulator, resistant microbe consortium, and nano iron complex. Environmental Pollution 257, 113558. Available at https://doi.org/10.1016/j.envpol.2019.113558.

Wang, G.-J., Liu, W.-L., Zhao, B., Shen, Y.-M., Wang, M.-L., Li, X.-S., 2022. Regional distribution and causes of global mine tailings dam failures. Metals 12 (6), 905. Available at https://doi.org/10.3390/met12060905.

Wen, H., Zhu, C., Zhang, Y., Cloquet, C., Fan, H., Fu, S., 2016. Zn/Cd ratios and cadmium isotope evidence for the classification of lead-zinc deposits. Scientific Reports 6 (1), 25273. Available at https://doi.org/10.1038/srep25273.

Werner, T.T., Mudd, G.M., Schipper, A.M., Huijbregts, M.A.J., Taneja, L., Northey, S.A., 2020. Global-scale remote sensing of mine areas and analysis of factors explaining their extent. Global Environmental Change 60, 102007. Available at https://doi.org/10.1016/j.gloenvcha.2019.102007, https://www.elsevier.com/inca/publications/store/3/0/4/2/5.

Wu, B., Wang, Z., Peng, D., Wang, Y., He, T., Tang, H., et al., 2020. Removal and recovery of heavy metals from soil with sodium alginate coated FeSSi nanocomposites in a leaching process. Journal of Hazardous Materials 398, 122732. Available at https://doi.org/10.1016/j.jhazmat.2020.122732.

Wójcik, M., Gonnelli, C., Selvi, F., Dresler, S., Rostański, A., Vangronsveld, J., 2017. Metallophytes of serpentine and calamine soils—their unique ecophysiology and potential for phytoremediation. Advances in Botanical Research 83, 1–42. Available at https://doi.org/10.1016/bs.abr.2016.12.002, http://www.elsevier.com/wps/find/bookdescription.cws_home/675801/description#description.

Xiao, T., Guha, J., Boyle, D., Liu, C.Q., Chen, J., 2004. Environmental concerns related to high thallium levels in soils and thallium uptake by plants in southwest Guizhou, China. Science of the Total Environment 318 (1–3), 223–244. Available at https://doi.org/10.1016/S0048–9697(03)00448-0, www.elsevier.com/locate/scitotenv.

Xiao, X., Zhang, J., Wang, H., Han, X., Ma, J., Ma, Y., et al., 2020. Distribution and health risk assessment of potentially toxic elements in soils around coal industrial areas: a global meta-analysis. Science of the Total Environment 713, 135292. Available at https://doi.org/10.1016/j.scitotenv.2019.135292.

Zarasvandi, A., Poursheikhi, E., Saki, A., 2021. Sulfide minerals and fluid chemistry of Zn–Pb deposits in central Sanandaj–Sirjan Zone, Iran. Iranian Journal of Science and Technology, Transactions A: Science 45 (6), 2001–2020. Available at https://doi.org/10.1007/s40995-021-01181-6.

Zhang, H., Xu, Y., Kanyerere, T., Wang, Y.S., Sun, M., 2022. Washing reagents for remediating heavy-metal-contaminated soil: a review. Frontiers in Earth Science 10, 901570. Available at https://doi.org/10.3389/feart.2022.901570, https://www.frontiersin.org/journals/earth-science.

Zhan, HuiYing, Jiang, YuFeng, Yuan, Jianmei, Hu, XueFei, Nartey, Obemah David, Wang, BeiLei, 2014. Trace metal pollution in soil and wild plants from lead–zinc smelting areas in Huixian County, Northwest China. Journal of Geochemical Exploration 147, 182–188. Available at 10.1016/j.gexplo.2014.10.007.

Zheng, X.-J., Li, Q., Peng, H., Zhang, J.-X., Chen, W.-J., Zhou, B.-C., et al., 2022. Remediation of heavy metal-contaminated soils with soil washing: a review. Sustainability 14 (20), 13058. Available at https://doi.org/10.3390/su142013058.

Zhou, L., Zheng, Y., Duan, X., Meng, Y., Yu, P.-P., Li, Z., et al., 2023. Carbonate-hosted Pb-Zn deposits in China: a review of the geological characteristics and genesis. In: Andrew, C.J., Hitzman, M.W., Stanley, G. (Eds.), Irish-Type Deposits Around the World. Irish Association for Economic Geology, Dublin, pp. 557–594. Available at https://doi.org/10.61153/EYLY2924.

CHAPTER THREE

Phytomining via phytoremediation

3.1 Introduction

The increasing intensity of human activities through the intensification of industrial production (especially mining) and agriculture has led to increasing soil pollution. In recent decades, phytoremediation has emerged as a novel technology that can help reduce contamination to recommended levels. Phytoremediation is a cost-effective and environmentally sound remediation method that utilizes plants and their associated microbes to remove organic and nonorganic pollutants from solid, gaseous, and liquid substrates (Pilon-Smits, 2005). Solid substrates are most commonly remediated, including mine tailings, industrial sites, and agricultural land. The water systems that are usually contaminated with high concentrations of pollutants are industrial effluents, mine drainage, agricultural effluents, municipal wastewater, sewage, etc., while the gaseous compounds that can be neutralized by plants are CO_2, NO_x, SO_2, etc. (Singh & Verma, 2007). Depending on how the remediation is carried out, we can distinguish several technologies: (1) phytoextraction, in which the pollutants are extracted from the soil and accumulated in the above ground tissues of the plants, mainly in the leaves; (2) phytostabilization, which refers to the stabilization of the pollutants in the soil or in the plant roots, preventing their further leaching into the surrounding systems; (3) rhizofiltration, where aquatic plants absorb and accumulate pollutants from the water; (4) phytodegradation/rhizodegradation refers to the degradation of organic pollutants with the help of plants or soil microbes; (5) phytovolatilization refers to the degradation of pollutants with the help of plants, releasing degradation products in volatile form. In addition to reducing the concentration of pollutants in the substrate, modern remediation includes additional ecosystem services and closes the entire cycle with the realization of certain economic benefits. This is the case with phytomining, that is, the use of plants to extract metals from the soil according to the principles of phytoextraction, one of the most commonly used phytoremediation

Metallophytes in Mineral Mining
DOI: https://doi.org/10.1016/B978-0-443-30110-0.00002-7

techniques. This not only ensures mining with minimal impact on the environment but also improves its properties. As this process involves the cultivation of plants according to the principles of agricultural production, this technique is more appropriately referred to as agromining.

3.2 What is phytomining?

Phytomining is the use of the hyperaccumulator plants to mine metals and metalloids in a process that, in addition to remediating the area, brings economic benefits as the plants are used as a Ni-ore (Chaney, 1983). Since this process, especially if it is to bring economic benefits, involves the cultivation of crops and the use of agricultural measures, it is more accurately referred to as agromining (Morel, 2013). The basis of both processes is the extraction of elements, with phytomining referring more to the potential of plants to extract elements from the soil, while agromining focuses on the technological process to improve crop yields and thus the amount of extracted elements (Chaney et al., 2021). This process, which was originally introduced for the processing of mine waste (Chaney, 1983), is now increasingly applied to naturally mineralized soils, especially ultramafic soils, and involves the implementation of various agronomic measures. After harvesting and incineration, the bio-ore obtained from the plant biomass is the source for the recovery of the metals from their salts (van der Ent et al., 2015). Although it is theoretically possible to work with several different elements (Zn, Mn, Cd, Ni, rare earth elements (REEs), etc.) in this process, previous experience has shown the best results with the recovery of Ni, mainly due to the high number of hyperaccumulators of this element (>500) (Reeves et al., 2018). Agromining studies started in the Mediterranean region with different *Odontarrhena* species (Bani et al., 2015a, 2015b; Li et al., 2003a, 2003b) and continued in the tropical region in Malaysia, especially with the species *Rinorea bengalensis* (Wall.) Kuntze (Violaceae) and *Phyllanthus rufuschaneyi* Welzen, R.W. Bouman & Ent (Phyllantaceae).

3.3 The key steps for establishing a phytomining system

3.3.1 Selection of a suitable land area

The most important criteria for a feasible phytomining process include the selection of plant species and the cultivation area. Due to the high concentrations of total Ni, ultramafic areas are considered the most promising for mining this element. There are two approaches to phytomining, which differ in the type of Ni-rich sites used: (1) phytomining in mined or degraded areas (tailing or areas close to smelters) and (2) agromining on agricultural land with low productivity (van der Ent et al., 2015). It has been shown that flat agricultural land with low productivity is unsuitable for crop production due to excessive Ni concentrations and poor nutritional status (lack of N, P, and K), making agromining cultivation economically more favorable compared to conventional agriculture. However, the most important factor in the selection of cultivation areas is a sufficiently high concentration of available Ni. The hyperaccumulators of Ni have a great potential for the uptake of elements, but high concentrations in the soil are necessary to make agromining efficient. Over time, the pool of labile Ni has been depleted by plant uptake, which has a positive effect on soil properties. While there are no results from studies that have been replicated often enough to claim with certainty, it is hypothesized that the mining of Ni could be sustainable for up to 50 years (Nkrumah et al., 2016).

3.3.2 Selection of appropriate plant species

There are several criteria for selecting a suitable plant species for the most efficient agromining process. The most important is a high potential for the uptake of elements and their translocation and accumulation in aboveground tissues. For Ni, the most commonly mined element, the threshold for hyperaccumulation has been set at 1000 $\mu g\ g^{-1}$ (van der Ent et al., 2013), with the best results obtained with hypernickelophores (10,000 $\mu g\ g^{-1}$, or 1%) as better candidates for metal crops (Nkrumah et al., 2021a). Plant species known as hypernickelophores are species from the genera *Bornmuellera* (*B. baldaccii* (Degen) Heywood, B. *dieckii* Degen, *B. tymphaea* (Hausskn.) Hausskn., *B. emarginata* (Boiss.) Rešetnik; Reeves et al., 1980; Paçarizi et al., 2020), *Odontarrhena* (as *O. baldaccii* (Vierh. ex Nyár.) Španiel, *O. chalcidica* (Janka) Španiel, Al-Shehbaz, D.A.German & Marhold, *O. lesbiaca* Candargy, *O. muralis* (Waldst. & Kit.) (Kazakou et al., 2010; Jakovljević et al., 2022; Reeves et al., 1997), *Phyllanthus* (*P. balgooyi* Petra Hoffm. & A.J.M. Baker, P. *erythrotrichus* C.B. Rob., *P. rufuschaneyi*, *P. securinegoides* (Quimado et al., 2015; van der Ent et al., 2022), then *Berkheya coddii* Roessler (Morrey et al., 1989) and *Rinorea bengalensis* (Datta et al., 2015). Additional criteria correspond to the basic requirements for successful phytoextraction and relate to a sufficiently high yield and high above ground biomass as well as indigeneity, that is, adaptation to local environmental conditions and reduction of the risk of possible invasiveness. Ease of germination and harvest would add value to this process, as would the perennial nature of the species, considering that planting is a one-time expense (Nkrumah et al., 2021a).

3.3.3 Planting technology

Previous studies on agromining have shown a number of practices that have a positive impact on the results obtained, both in terms of increased accumulation of Ni from the soil and in terms of an increase in plant biomass, all of which have an impact on Ni yield (Nkrumah et al., 2021a). Before planting, tillage is recommended to allow the roots to reach the deeper soil layers and soil amendments. Planting density has been shown to be of great importance in maximizing results, although it varies by species. In the studies on *O. chalcidica*, planting four plants per square meter was recommended to achieve the best results, as three times the yield was obtained this way compared to planting six plants per square meter (Bani et al., 2015b). The higher yield was also observed in the absence of other species (mainly weeds) due to reduced competition for water and essential elements, which can lead to increased mortality in certain plots (up to 60%; Bani et al., 2015b), while the use of rhizobacteria improved Ni accumulation by increasing elemental availability (Durand et al., 2016). Co-cropping is also one of the techniques considered potentially useful for agromining, with stress reduction, higher biomass, and more efficient extraction observed in some studies (Cid et al., 2016, 2020; Jiang et al., 2010, 2015). Certain combinations of species not only show a positive effect on growth but also lead to lower extraction of Cu and Zn, as observed for *Zea mays* L. and *Sedum alfredii* Hance. This is particularly positive for *Z. mays*, where the risk of the metal entering the food chain is lower (Xiaomei et al., 2005). The use of co-cropping in Ni extraction was also of variable success, with positive results (increase in plant biomass and Ni yield) for the combination of *O. chalcidica* and *Vicia sativa* L. (Saad et al., 2021), whereas *Lupinus albus* L. reduced Ni extraction in the shoots of the same species (Jiang et al., 2015).

3.3.4 Harvesting strategies

The planting and harvesting of plant species for agromining purposes is highly dependent on the plants and site characteristics and is therefore species- and site-specific. The practice of planting *Odontarrhena* species, which are the most intensively used for agromining, is determined by the climatic characteristics of the Mediterranean region, where these species are predominantly distributed. Therefore, the planting of seedlings in the field is usually prepared for the autumn, when the dry season is over. During the dry season (hot and dry summer), the seedlings are prepared using irrigation water. The following summer, during the flowering period, in mid-June, the plant shoots are cut and dried so that they are ready for the next production step (Kidd et al., 2018). The middle of the flowering period was also considered optimal for harvesting *B. emarginata*, which is used for agromining in Spain alongside *O. chalcidica*. In addition, not only the leaves of *B. emarginata*, especially those of the flowering stems, but also flowers and fruits contributed strongly to the Ni yield (Kidd et al., 2018). Unlike in the Mediterranean region, woody species such as *R. bengalensis* and species from the genus *Phyllanthus* are used for agromining in the tropics, where the lack of seasonal leaf fall requires harvesting several times a year to collect the leaf litter with a lot of accumulated Ni (Nkrumah et al., 2021b).

3.3.5 Postharvest treatment of biomass and marketing

The treatment of the harvested biomass and its transformation into the product of interest to the market is an important step in agromining, where the Ni is recovered through thermal and chemical processes. Before further processing, the harvested biomass is dried in the air and in the sun to obtain biomass with a Ni content of up to 2% wt. In addition to Ni, the biomass is also rich in macroelements such as K, Ca and Mg, organic carbon and various impurities (Zhang et al., 2014). In the next step, the biomass is ashed to remove impurities and organic matter and, most importantly, to concentrate the elements (Barbaroux et al., 2012; Cassayre et al., 2020; Zhang et al., 2014). This process generates a considerable amount of energy that can be utilized further. In addition to the concentration of the elements, the ashing of the biomass also significantly reduces the size of the plants required to extract the elements and the cost of transporting the intermediate products. Previous studies on Brassicaceae species have revealed several aspects whose variations strongly influence the utilization efficiency. One of these is the ashing temperature, which should be high enough to break down organic matter, but not so high that metals are lost through volatilization. It is also important to determine the optimal incineration time to burn all organic matter and ensure optimal utilization of the furnace (Laubie et al., 2021).

In addition to this treatment with biomass incineration, where the Ni is recovered from the ash by leaching in a hydrometallurgical process and the desired products are obtained by crystallization or precipitation, there are several different treatments for Ni extraction. One of the methods involves carbonization of the biomass, which produces a catalyst as an end product. Due to the intensive use of catalysts in the chemical industry, this process is economically feasible, but is not yet widely used. Nickel can also be extracted directly from the biomass, either by high temperatures or by leaching in a process that produces a number of undesirable by-products (Laubie et al., 2021).

3.3.6 The economics of the agromining process

In view of the considerable decline in conventional ore grades, the development of alternative measures to achieve sufficient metal concentrations is imperative. This is particularly true for some of the elements that are abundant in ultramafic soils, such as Ni and Co, but are now considered strategic metals or even critical raw materials (Laubie et al., 2021). In fact, the previous practice of Ni production was mainly related to sulfide ores and laterites, which are mainly found in Indonesia, the Philippines, New Caledonia, Brazil, and so on. With the intensive exploitation of sulfide ores and the significant decline in deposits, laterites became the main source of Ni, although this production process has proven to be very expensive and unsuccessful in recovering Co (Battle et al., 2013). At the same time, the demand for Ni increased dramatically as it is used for the production of Li-ion batteries and has been in demand for decades in the metallurgy, construction, chemical, and petrochemical industries. Considering the prices of Ni and its products, Ni salts (Fig. 3.1) are considered the most important products containing production raw materials of high purity (Laubie et al., 2021). At the same time, the use of agromining to recover Ni is the most promising, as this element has been shown to be the most commonly hyperaccumulated, with more than 520 species found to have more than 1000 $\mu g\ g^{-1}$ of Ni in the leaves (Reeves et al., 2018). The earlier practice of recovering metals was mainly related to ultramafics and hyperaccumulators of Ni, and, after successful tests under laboratory conditions, was successfully transferred to the fields, where it has been applied in recent decades with some modifications of the technological process (Laubie et al., 2021). Selected agronomic practices proved to be highly important for maximizing the yield of the metal crops (Nkrumah et al., 2021a), and their intensive use on *O. chalcidica* resulted in an increase in

Figure 3.1 Ammonium Ni-sulfate hexahydrate, a salt synthesized in the agromining process.

Ni yield from 100 to 200 kg ha^{-1} (Nkrumah et al., 2016), whereas the estimated Ni yield achieved by applying agrotechnical measures to the species *P. rufuschaneyi* in Sabah was 250 kg ha^{-1} (Nkrumah et al., 2019b).

In the economics of the agromining process, two main groups of activities are distinguished: cultivation and conversion. Cultivation involves renting a plot of land, hiring labor, preparing the land, buying seeds, putting fertilizers, and providing irrigation, where the involvement of local farmers in planting and harvesting significantly reduces costs and facilitates the process not only due to their knowledge of the area and the local plant species used for agromining but also due to their familiarity with local legislation. However, despite all the economic benefits, it should not be forgotten that harvesting most species is still a manual process that is both tedious and expensive. Conversion, on the other hand, encompasses all processes in which products of various metals are extracted from biomass, the price of which largely determines the feasibility of the entire agromining process. In an ideal agromining process, the energy released during combustion should also be utilized, which can provide some benefit, although this process depends on many variables, especially the proximity of the process facilities (Kidd et al., 2018).

Given the complexity of the process and all the costs associated with preparing the plant material and processing the biomass, it is not easy to determine the potential economic gain, which is known to be highly significant on the plant species used, the area under cultivation, and the applied process. Nevertheless, some estimates can be made. It is assumed that the net gain of the agromining process, which would include the species *O. chalcidica*, is about 80% of the Ni value (Bani et al., 2015a), while this percentage is estimated at 75% in Malaysia (Kidd et al., 2018).

3.4 Pioneering phytomining

The concept of phytomining was initially referred to waste materials and was introduced by Chaney in the 1980s (Chaney, 1983; Chaney et al., 1981). The beginnings of the application of phytomining relate to the USA and to areas heavily contaminated with Cd, as negative health effects have been recognized. However, after the initial funding of the program, the lack of a corresponding market led to the reallocation of funds within a very short time, whereupon the research was discontinued (Chaney et al., 2021). With the realization that phytomining Ni could be economically feasible, research in Europe was redirected to species of the genus *Odontarrhena* and in particular to the species *O. chalcidica*. Attempts at Ni-phytomining have also been made in the USA, in connection with the species *Streptanthus polygaloides* A. Gray from ultramafics in California. Although this species belongs to the hypernickelophores and contains more than 1% Ni, its exploitation proved to be unfeasible because the leaves do not reach sufficient mass (Reeves et al., 1981).

In addition to *O. chalcidica*, other *Odontarrhena* species from the Mediterranean region have also been studied in Europe (*O. muralis* s.l., *O. serpyllifolia* (Desf.) Jord. & Fourr.), which are known to grow in ultramafic areas with excessive Ni concentrations in the soil, emphasizing the soil–plant system and the most optimal conditions for the uptake of the element (Echevarria et al., 1998; Gérard et al., 2000). What is interesting about these species is that the highest efficiency in Ni extraction is achieved at higher pH values (Kukier et al., 2004), so that studies relying on chelates to increase bioavailability, which was very common in the early days of phytomining, showed unexpectedly poor results. Research in the Balkan Peninsula extended from *Odontarrhena* to *Bornmuellera* species, mainly *B. tymphaea* and *B. emarginata*, also known as *Leptoplax emarginata* (Boiss.) O. E. Schulz (Chardot et al., 2005; Kyrkas et al., 2023, 2024; Ly et al., 2025), all known as hypernickelophore (Jakovljević et al, 2022).

After a successful phase of testing and application of agromining in temperate regions, mainly in Greece, Albania, Spain, and Austria (Bani et al., 2021; Kidd et al., 2018; Kyrkas et al., 2023; Ly et al., 2025; Pardo et al., 2018;

Rosenkranz et al., 2019), agromining studies have been extended to tropical regions, with trial fields in Sabah (Malaysia) (Nkrumah et al., 2019b). Initial studies were also initiated in Mexico, where considerably high Ni concentrations were detected using XRF, as in *Blepharidium guatemalense* Standl. with more than 2% Ni in the leaves, opening up possibilities for the development of agromining in this area (Nkrumah et al., 2021a).

3.5 Hyperaccumulators

Plants respond differently to excessive elemental concentrations in the substrate, and depending on this reaction, three main groups can be distinguished: excluders, indicators, and accumulators (Baker, 1981). While the element concentrations in the shoots of excluders remain constant despite the excessive concentrations in the soil (whereby the elements are not taken up at all or are retained in the root), those in the shoots of indicators are regulated by the concentrations in the substrate and reflect their level accordingly. Plants that concentrate elements in their shoots (both via high and low concentrations in the substrate) are called accumulators. An extreme group of accumulators are the hyperaccumulators, which are plants that are able to take up elements in their foliar tissue in concentrations 2–3 orders of magnitude higher than nonaccumulators without showing symptoms of toxicity. These properties make hyperaccumulators indispensable for the agromining process. According to Reeves et al. (2018), this phenomenon has so far been recognized in 721 species belonging to 130 genera and 52 families, being most common in the families Brassicaceae (83 taxa), which are mainly distributed in temperate and Mediterranean regions, and Phyllanthaceae (59 taxa), which are associated with tropical regions (Reeves et al., 2021). To be considered a hyperaccumulator, plant species must fulfill several criteria: (1) a foliar concentration of the accumulated element above the hyperaccumulation threshold and (2) higher elemental concentrations in the leaves compared to the roots and extractable in the soil (shoot-to-root ratio > 1 and bioconcentration factor > 1, but more often >50) (van der Ent et al., 2013). Importantly, all these criteria must be met in naturally growing plants, while observations on experimentally cultivated plants per se are not sufficient to consider species as hyperaccumulators. According to the different requirements of plants for the various elements, the threshold values for hyperaccumulation are different and range from 100 $\mu g\ g^{-1}$ for

Cd, Tl, and Se, 300 μg g^{-1} for Co, Cr, and Cu, 1000 μg g^{-1} for As, Ni, Pb, and rare earth elements, 3000 μg g^{-1} for Zn, and 10,000 μg g^{-1} for Mn (van der Ent et al., 2013).

3.5.1 Nickel hyperaccumulator plants

Previous studies on the phenomenon of hyperaccumulation in plant species have shown that most of these species hyperaccumulate Ni (532 taxa; Reeves et al., 2018). The predominance of Ni hyperaccumulators is due to the early recognition of their specificity and the analysis of flora developed on ultramafic geological substrates rich in Ni, as well as the availability of rapid tests for detecting high foliar Ni concentrations. This test, which is mainly used in the field, involves the use of paper treated with dimethylglyoxime, which turns pink when pressed onto the leaves containing Ni (Baker et al., 1992). To date, most hyperaccumulators of Ni are members of the Brassicaceae family, being particularly abundant in the genera *Odontarrhena* and *Noccaea* (Drozdova et al., 2019). In the study of almost 170 *Alyssum* taxa known at that time, 48 hyperaccumulators of Ni alone were discovered. These taxa, previously representatives of the *Odontarrhena* section, were later recognized as a separate genus under the same name.

Particularly important from an agromining perspective are hypernickelophores, which are plants with more than 1% Ni in their foliar tissue (Jaffré et al., 2013), which mainly belong to the genera *Bornmuellera, Odontarrhena, Phyllanthus, Berkheya*, and *Rinorea* (Nkrumah et al., 2021a). Some of the highest Ni concentrations in leaves were found in *O. lesbiaca* (23,650 μg g^{-1}; Kazakou et al., 2010), *Noccaea boeotica* F. K. Mey. (23,400 μg g^{-1}; Reeves and Brooks, 1983), *B. emarginata* (34.400 μg g^{-1}; Reeves et al., 1980), *Phyllanthus* cf. *securenegoides* (23,300 μg g^{-1}; van der Ent et al., 2015), and *B. coddii* (30,000 μg g^{-1}; Rue et al., 2020). Latex is also an important source of Ni and contains 10% of the Ni in *Pycnandra acuminata* (Pierre ex Baill.) Swenson & Munzinger (Jaffré et al., 1976).

In terms of distribution, two main groups of Ni hyperaccumulators can be distinguished: (1) those from the temperate region, centered on the Mediterranean, the Balkan Peninsula, and Turkey and (2) tropical and subtropical regions, particularly Cuba, Brazil, and New Caledonia. While representatives of the Brassicaceae family dominate in the temperate regions, plants from the Phyllanthaceae, Violaceae, Rubiaceae, Euphorbiaceae, and Buxaceae families dominate in the tropical and subtropical regions (Reeves et al., 2021).

3.5.2 Hyperaccumulators of zinc and cadmium

Zinc is one of the essential elements for plants, a cofactor for more than 300 proteins and the only element that occurs in all six enzyme groups. It also plays an important role in the detoxification of plants, the stabilization of membranes and the metabolism of carbohydrates (López-Millán et al., 2005; Vaculík et al., 2025). Although Zn deficiency is more frequently reported, there are still plants with foliar Zn concentrations above the proposed hyperaccumulation threshold (3000 $\mu g\ g^{-1}$; van der Ent et al., 2013). To date, there are more than 30 taxa with this potential (Vaculík et al., 2025). Similar to the hyperaccumulators of Ni, most of them are representatives of the Brassicaceae family, belonging to the genera *Noccaea*, *Arabis*, and *Arabidopsis*. The highest Zn concentrations found in their taxa are 53,900 $\mu g\ g^{-1}$ in *Arabidopsis halleri* (L.) O'Kane & Al-Shehbaz (Stein et al., 2017), 53,450 $\mu g\ g^{-1}$ in *Noccaea caerulescens* (J. Presl & C. Presl) F. K. Mey. (Reeves et al., 2001), and 20,700 $\mu g\ g^{-1}$ in *Arabis paniculata* Franch. (Tang et al., 2009).

Although these plants occur mainly on calamine soils (rich in Cd, Zn, and Pb), some of the Zn hyperaccumulators were found on Zn-poor soils. The above-mentioned Zn concentration in *A. halleri* was found in plants growing on soils with less than 400 $\mu g\ g^{-1}$ Zn (Stein et al., 2017), whereas almost 9000 $\mu g\ g^{-1}$ Zn was detected in the leaves of *N. caerulescens* in areas with slightly more than 100 $\mu g\ g^{-1}$ Zn in the soil (Reeves et al., 2001). Similar findings were made for *Dichapetalum gelonioides* subsp. *pilosum* Leenh. and *D. gelonioides* subsp. *sumatranum* Leenh., which grow exclusively on Zn-poor soils, but whose foliar Zn concentrations reach above 26,000 and above 15,000 $\mu g\ g^{-1}$, respectively (Baker et al., 1992). *Cardamine waldsteinii* Dyer is one of the species in which high foliar concentrations of Zn (2210 $\mu g\ g^{-1}$) were found in a Zn-poor soil (66 $\mu g\ g^{-1}$ pseudo-total Zn) (Jakovljević et al., 2023).

In addition to *A. halleri* and *N. caerulescens*, model species for studying of Zn hyperaccumulation are also two *Sedum* species, *S. alfredii* and *S. plumbizincicola* X. H. Guo & S. B. Zhou ex L. H. Wu, which can also hyperaccumulate Cd. For *S. plumbizincicola*, both field and experimental studies confirmed the ability of the plants to take up these elements above the hyperaccumulation threshold (300 $\mu g\ g^{-1}$ and 3000 $\mu g\ g^{-1}$, respectively; van der Ent et al., 2013). In soil with extractable Cd and Zn concentrations of 7.6 $\mu g\ g^{-1}$ and 160 $\mu g\ g^{-1}$, respectively, foliar concentrations of these elements reached 2200 $\mu g\ g^{-1}$ and 16,400 $\mu g\ g^{-1}$, respectively. In addition, Cd concentrations in leaves were found to be twice as high as in stems, whereas Zn concentrations fluctuated within a very narrow range, with foliar concentrations only slightly lower or even equal to those in stems (Hu et al., 2014). Slightly lower concentrations were found in *S. alfredii*, with foliar concentrations of Zn up to 12,000 $\mu g\ g^{-1}$ and Cd up to 1400 $\mu g\ g^{-1}$ (Deng et al., 2007).

Hyperaccumulation of Cd is a relatively rare phenomenon that has been observed in seven taxa so far (Reeves et al., 2018). The threshold value for hyperaccumulation was set at 100 $\mu g\ g^{-1}$ (van der Ent et al., 2013) and was mostly exceeded in plants growing on the calamine soil. These were predominantly representatives of the genus *Noccaea* (*N. praecox* and *N. caerulescens*), in which Cd concentrations on the leaves reached more than 1000 $\mu g\ g^{-1}$, albeit with strong fluctuations within and between populations (Reeves et al., 2021). Leaf concentrations of more than 3600 $\mu g\ g^{-1}$ were found in *A. halleri* from a metal-rich site in Poland (Stein et al., 2017).

3.5.3 Gold hyperaccumulator plants

Due to its rare occurrence and economic value, gold is considered a precious metal together with the elements of the platinum group (iridium, palladium, platinum and rhodium) (van der Ent et al., 2021). Although it is not an essential element for the plant, it is easily taken up in soluble form and translocated to the aerial parts. However, the concentrations found in plant species, even at auriferous sites, are far from those commonly observed in accumulating and hyperaccumulating species. Previous data on Au concentrations in plant shoots were somewhat misleading, as for Au accumulation in ash of *Equisetum palustre* L., noted to be above 600 $\mu g\ g^{-1}$ (Babička, 1943). However, more recent and more precise data on Au accumulation indicate that *E. palustre* is only an indicator of auriferous sites and not an accumulating species. Relatively high Au concentrations were also found in *Tsuga heterophylla* (Raf.) Sarg. (0.3 $\mu g\ g^{-1}$), but still far from the notional hyperaccumulation threshold of 1 $\mu g\ g^{-1}$ (Anderson et al., 2003; van der Ent et al., 2021).

3.5.4 Hyperaccumulators of rare earth elements

Increasing commercial interest in REEs has fueled research into the accumulation of these elements in plant species. REEs are a group of 17 chemical elements—15 lanthanides, yttrium, and scandium—and are usually categorized into light rare earths (LREEs) and heavy rare earths (HREEs) (Khan et al., 2017). Although concentrations of 1000 $\mu g\ g^{-1}$ are considered the threshold for hyperaccumulation of REEs (Baker and Brooks, 1989), the final value has not yet been established, with evidence suggesting that it can be lowered to 100 μg g^{-1} (van der Ent et al., 2021). Only a few species have been recognized as REE hyperaccumulators, mainly due to their low concentrations in soil combined with low availability (Purwadi et al., 2025).

Most REE hyperaccumulators were found in the pteridophytes, both in contaminated mine sites and in uncontaminated natural sites. Some of the highest REE concentrations in leaves were found in *Dicranopteris linearis* (Burm.f.) Underw. (7000 μg g^{-1}; Shan et al., 2003), *Pronephrium simplex* (Hook.) Holttum (3000 μg g^{-1}; Lai et al., 2006), and *Blechnum orientale* L. (1000 μg g^{-1}; Xiao et al., 2003).

In addition to ferns, hyperaccumulation of REE was also observed in angiosperms, such as *Phytolacca americana* L. (1040 μg g^{-1}; Liu et al., 2021). Experimental analysis revealed the dominant pathways of REEs, which differ for light and heavy REEs. While the light REEs were taken up via Ca^{2+} channels, Al^{3+} channels predominated in the heavy REEs (Yuan et al., 2017). Members of the Juglandaceae family also accumulate REEs above the nominal hyperaccumulation threshold, with 2300 μg g^{-1} REE in *Carya glabra* (Mill.) Sweet (Robinson et al., 1938) and 1350 μg g^{-1} in *Carya tomentosa* (Lam.) Nutt. (Thomas, 1975), but also *Juglans nigra* L., which show hyperaccumulation of yttrium in herbarium specimens (Goudard et al., 2024).

3.5.5 Arsenic hyperaccumulator plant

Arsenic is a highly toxic metalloid, nonessential for plants, whose excessive concentrations in soil have been caused by both natural (mineralogical) and anthropogenic sources (Gonzaga et al., 2006). Similar to REEs, some of the highest As concentrations have been found in ferns, *Pteris vittata* L. being the best known hyperaccumulator of this element, with up to 5000 μg g^{-1} As in the fronds of plants grown in a site contaminated with chromated copper arsenate, whereas in plants experimentally grown for 6 weeks in soil treated with 1500 μg g^{-1} As, concentrations in the fronds reached more than 22,000 μg g^{-1} (Ma et al., 2001). Concentrations above the hyperaccumulation threshold were also found in another fern, *Pityrogramma calomelanos* (L.) Link, with 8350 μg g^{-1} of As in fronds (Francesconi et al., 2002).

Recent studies on the hyperaccumulation of elements in plants have revealed some new hyperaccumulators of arsenic, such as in the area of

Allchar, an abdandoned As-Tl mine in North Macedonia (Jakovljević et al., 2024). As concentrations of more than 1000 μg g^{-1}, together with TF >1, were found in individuals of *Alyssum kavadarcense* Micevski (1330 μg g^{-1}), *Minuartia verna* (L.) Hiern (1340 μg g^{-1}), and *Silene vulgaris* (Moench) Garcke (2780 μg g^{-1}).

3.5.6 Thallium hyperaccumulator plants

Thallium is a highly toxic metal, of which only a few species are known to be hyperaccumulators. The interest in phytomining of Tl started with the discovery of above 500 μg g^{-1} of this element in foliar tissue of *Iberis intermedia* Guers. and an estimated profit of US$1000 per year (Robinson and Anderson, 2021). Besides *I. intermedia*, not many species were known to be hyperaccumulators of this element, with concentrations above 100 μg g^{-1} as a hyperaccumulation threshold (van der Ent et al., 2013) also found in *Biscutella laevigata* L. (Corzo Remigio et al., 2022) and *Silene latifolia* Poir. (Jakovljević et al., 2025). The Tl concentration of 79,200 μg g^{-1} found in the leaves of *S. latifolia* from Allchar, an abandoned As-Tl mine, is the highest known in nature (Jakovljević et al., 2025). In addition to *Silene*, concentrations above the hyperaccumulation thresholds were also found in several species from Allchar, such as *M. verna* and the endemic Allchar species *A. kavadarcense*, *Centaurea leucomala* Bornm. and *Knautia caroli-rechingeri* Micevski (Bačeva Andonovska et al., 2021; Jakovljević et al., 2024).

3.6 Strategies to enhance the economics of phytomining

The selection of the most suitable plant species and the most suitable land are the first and most important steps to ensure economically viable agricultural production. However, similarly to classical agricultural practices, there are some measures that can be applied (fertilizers, herbicides, and pH adjustment) to improve the growth of the plants, their biomass and therefore the yield of the mined element. Achieving optimal growth conditions in terms of nutrients, moisture, and plant density is very important in order to maximize the efficiency of extraction with the various measures proposed for the uptake of the different elements.

As this is the element with which the practice of agromining started and later developed, most agronomic measures have been developed to maximize Ni yield, both in the temperate region with *Odontarrhena* species and in the tropics with *Rinorea bengalensis* and *Phyllanthus rufuschaneyi* (Bani et al., 2015a; Kidd et al., 2018; Nkrumah et al., 2019a, 2019b). Metalliferous areas, both of natural and anthropogenic origin, are characterized by a deficiency or even absence of macroelements, especially N, P, and K. The addition of fertilizers rich in these elements proves to be extremely effective in improving plant growth and yield (Álvarez-López et al., 2016; Nkrumah et al., 2019a). The application of N-, P- and K-based fertilizers on the ultramafic plots with *O. chalcidica* led to a doubling of plant biomass (Bani et al., 2015a), whereas the use of Ca and herbicides in addition to N, P, and K led to a 10-fold increase in yield in this species (Bani et al., 2015b). The lack of calcium is also a common feature of ultramafic soils with a characteristically low Ca/Mg ratio. Considering that increasing the Ca/Mg ratio has been shown to reduce metal stress in plants (Galey et al., 2017), the addition of Ca has a strong effect on plant yield. Plants compensate for Ca deficiency by increasing the uptake of this element, whose already low concentrations decrease rapidly in this way, with an estimated annual depletion of 2%. It has been shown that calcium taken up by plants and translocated to the leaves increases the tolerance of plants to Ni toxicity, but has no effect on Ni accumulation (Chaney et al., 2008). Compared to inorganic fertilizers, organic fertilizers applied in the form of municipal solid waste did not show significant effects on plant yield, mainly due to the lower bioavailability of Ni (Nkrumah et al., 2021a). On the contrary, an increase in biomass was observed with the addition of the two main groups of fertilizers, which is particularly important for the cultivation of the hyperaccumulator plant species on waste material. The addition of sulfur also proved to be extremely beneficial in agromining studies, especially for members of the Brassicaceae family, which require this element for the metabolism of glucosinolates (Jankowski et al., 2015). The role of glutathione has been recognized in Ni tolerance and its precursor, which is rich in S, has been shown to be involved in the hyperaccumulation of Ni in *Noccaea* species, with these two elements colocalized in vacuoles (Broadhurst et al., 2004). The addition of sulfur has also been associated with an improvement in Ni availability, as shown in *Berkheya coddii*, both by lowering pH and increasing available S (Robinson et al., 1999). Soil acidity is one of the most important factors for the uptake of elements and is usually maximized at lower values. The highest uptake of Ni in ultramafic soils was demonstrated at pH values of 6.5, with an increase observed in the range of 5–6.5 (Nkrumah et al., 2021a).

In tropical regions (Malaysia), where this type of research is still in its infancy, agrotechnical measures have also been applied in the fields. Research here is mainly focused on the species *R. bengalensis* and *P. rufuschaneyi*, where large-scale studies have been conducted following successful trial studies under controlled conditions (Nkrumah et al., 2019b). Similar to temperate regions and trials with *Odontarrhena* species, inorganic fertilizers also improved biomass and Ni yield in these two species, with the exception of K supplied to *P. rufuschaneyi*, where both growth and Ni yield decreased. The same was observed for organic matter when added to *Phyllanthus* on a substrate such as cambisol. However, in some extreme substrate types with a high lack of organic matter, this supplementation could be necessary (Nkrumah et al., 2016).

The study of calamine soils rich in Cd, Pb, and Zn and their prevalent species made it possible to develop a practice that allows a significant increase in yield on this type of soil (Shen et al., 2011; Wu et al., 2021). *Sedum plumbizincicola* is a model species for studying the accumulation of Cd and Zn, considering that field and experimental studies have confirmed the plant's ability to take up these elements above the hyperaccumulation threshold (300 $\mu g\ g^{-1}$ and 3000 $\mu g\ g^{-1}$ respectively; van der Ent et al., 2013). In soils with extractable Cd and Zn concentrations of 7.6 $\mu g\ g^{-1}$ and 160 $\mu g\ g^{-1}$, foliar concentrations of these elements reached 2200 $\mu g\ g^{-1}$ and 16,400 $\mu g\ g^{-1}$, respectively. Furthermore, Cd concentrations in leaves were found to be twice as high as in stems, whereas Zn concentrations varied in a much narrower range, with foliar concentrations only slightly lower or even equal to those in stems (Hu et al., 2014). These high concentrations in the harvestable parts of *S. plumbizincicola* suggest that its use in the agromining process is possible. However, plant preferences implied the benefit of agronomic measures to improve potential outcomes. The optimal plant density that allows for maximum uptake of Cd and Zn is 15 cm × 15 cm, with uptake decreasing both when the area is increased and decreased (Wu et al., 2021). As a shade plant, no reduction in uptake was observed at irradiance greater than 50% (Li et al., 2010), whereas the optimum water storage capacity was estimated at 70% (Cui et al., 2009). This shade tolerance allows co-cropping with tall crops such as sorghum, maize, and wheat, and the efficiency of Cd and Zn extraction has been confirmed (Deng et al, 2016). However, caution is required as *Sedum* roots have increased Cd availability, leading to potential toxicity in the crops used (Zhao et al., 2011). The addition of nutrients can also strongly influence agromining results. In terms of nitrogen, previous studies have shown that the use of nitrate (NO^{3-}) compared to ammonium (NH^{4+}) provides more satisfactory results for biomass increase and Cd uptake and conversion into shoots. In addition, a high K content and a low P content were found to be favorable for the extraction of Cd in this species (Shen et al., 2011). The addition of organic matter was shown to greatly increase the bioavailability of Cd, by 2-fold when ground clover leaves and rice straw were added. Considering the preference for lower pH values, the most satisfactory results were obtained when treated with EDTA, which is known to increase the availability of elements. However, as this treatment increases the risk of toxicity, the use of chelates to increase bioavailability is generally not recommended. On the other hand, treatment with lime at a certain dose can increase the biomass and yield of Cd and Zn (Wu et al., 2021) while reducing the availability of other toxic elements from the substrate (Ren et al., 2013), although it can be assumed that the application of lime is not recommended due to its affinity for acidic environments. Due to its rapid growth, the plant can be harvested two to three times per year, with the stems remaining up to 5 cm tall, allowing the plant to reproduce (Li et al., 2009). The satisfactory results of using hyperaccumulator plant species suggested simultaneous cultivation with crops, as was the case with *S. plumbizincicola* and rice (Hou et al., 2019).

Phytoextraction can also benefit greatly from molecular knowledge in improving aspects of the physiological process: (1) mobilization of elements and uptake into plant roots; (2) transport of elements into root tissues with a reduction of sequestration in root vacuoles; (3) efficient loading of xylem and translocation of elements from root to shoot; and (4) unloading of xylem and distribution of elements in shoots (mainly in vacuoles). The identification of the main mechanisms underlying elemental accumulation is the most important step in genome editing technologies that will potentially have great benefits for phytoextraction processes and the economics of agromining.

Over the past 20 years, intensive research has been conducted to identify genes involved in the uptake and accumulation of elements (Charlier et al., 2015; Spielmann et al., 2020). Most of these studies related to *A. halleri* and *N. caerulescens*, hyperaccumulators of Zn and Cd, with two *Sedum* species (*S. alfredii* and *S. plumbizincicola*) recognized as new model species for the accumulation of these elements (Cao et al., 2014; Clauss and Koch, 2006; Li et al., 2018). Compared to nonaccumulator species, (hyper)accumulator plant species have been observed to secrete chelators that increase the solubility of metals in soil and facilitate uptake.

The next phase is mainly mediated by the proteins of the ZIP family, which are highly expressed in the roots and shoots and lead to an increased mobilization of elements from the root and transport into the shoot tissue (Krämer et al., 2007; Wu et al., 2009). For the ZIP transporter IRT1 (which is mainly involved in Fe uptake and in Zn, Ni, and Cd uptake due to its low selectivity), a strong effect on differences in Zn and Cd accumulation has been demonstrated (Halimaa et al., 2019). High expression of NAS (nicotianamine synthase) genes, which are responsible for translocation into the shoots, was also observed in these two model species, which is significantly higher than in nonaccumulating species (Liang et al., 2014). Similar results were found for not only HMA4 in *A. halleri*, a protein present in xylem loads (Hanikenne et al., 2008), but also for *N. caerulescens*, *S. alfredii*, and *S. plumbizincicola* (Peng et al., 2017; Van De Mortel et al., 2008; Zhang et al., 2016). Although HMA4 proteins were thought to play a role in the unloading and accumulation of elements in the shoots, it appears that these processes were mainly regulated by the MTP1 protein (metal tolerance protein 1), at least in *A. halleri* and *N. caerulescens* (Krämer, 2005; Milner and Kochian, 2008), whereas the heavy metal ATPase 1 (ATP1) was strongly expressed in the shoots of *S. plumbizincicola*, where it regulates the exclusion of elements from the chloroplast (Zhao et al., 2019).

The process of uptake and accumulation of Ni is somewhat similar to that of Zn and Cd, in terms of the role of IRT1 in elemental uptake and NA in Ni binding. In the prevention of root vacuolar sequestration, histidine showed to have a main role, whereas translocation of Ni from the root to the shoot was supported by carboxylic acids, as confirmed in *O. serpyllifolia* by the Ni-citrate complex (Alves et al., 2011). Most of the Ni was accumulated in the leaf vacuoles with ferroportin/iron-regulated transporters having a main role in sequestration (Morrissey et al., 2009). The uptake, translocation, and accumulation of other trace elements remained largely unknown, although main transporters group have been identified as MTP, ZIP, and NRAMP for manganese (Li et al., 2019) or those from SULTR family involved in selenite uptake (Barberon et al., 2008).

Positive effect in elemental uptake in plant species was observed in the presence of endophytic bacteria, as found for Ni translocation in *O. serpyllifolia* and *N. caerulescens* (Ma et al., 2011; Visioli et al., 2015) and Cd in *S. alfredii* (Wu et al., 2020). Benefits include improved protection against abiotic stress and other pathogens and availability, uptake, and transport of metals through the production of chelators and organic acids (Visioli et al., 2015).

3.7 Advantages and limitation of phytomining

Agromining, as a green mining technology, is mainly associated with areas that are mineralized, but not enough to start (or continue) conventional mining. So far, the most intensive studies on *Odontarrhena* species have been conducted in ultramafic areas where the Ni content is, on the one hand, insufficient to be economically exploited with classical mining methods and, on the other hand, so high that the soil should not be used for agricultural purposes. The lack of nutrients and the very often poor physical properties (thin, skeletal soil) also prevent these land from being used for conventional agriculture. The advantages of agromining in the utilization of such land are (1) the extraction of elements that would not be economical to extract using conventional mining methods; (2) in contrast to conventional mining, no additional environmental damage is caused; and (3) the intensive cultivation of plants leads to the enrichment of the substrate with organic substances and to the improvement of soil properties, making agromining not only a method of Ni-mining but also a method of soil remediation. The improvement of soil properties is even more intensive when considering the range of agrotechnical measures to increase plant growth and the yield of the mined element.

In addition to ultramafic areas with low productivity, significant benefits can be derived from the mining of waste material where elemental concentrations are also not high enough to ensure economic justification. It is estimated that from galvanic sludge containing 80 g kg^{-1} Ni, the species *O. chalcidica* could recover Ni in concentrations of 150 g kg^{-1} after incineration (Tognacchini et al., 2020).

Another advantage is the fact that a considerable number of countries are recognized as potentially good candidates for agromining. In Europe, in addition to countries where field studies already exist (such as Austria, Greece, Spain, and Albania), there are some countries such as Serbia, Bosnia, Turkey, and North Macedonia with significant ultramafic areas (Stevanovic et al., 2003), where agromining could replace agriculture on ultramafic soils with low fertility, bringing significant economic benefits. Furthermore, the extraction of nickel produces salts that have a higher market price compared to metallic Ni, which represents an additional economic advantage. Incineration of plant material generates a considerable amount of energy that can be reused in production processes or for heating.

At the same time, there are a number of factors that can reduce the efficiency of agromining. Most of these relate to the availability and uptake of elements by plants, factors that directly determine the efficiency of phytoextraction. In addition, antagonism in the uptake of some elements has also been observed. This is particularly true for Ni and Zn, as genes such as those of the ZIP family (zinc-regulated transporter iron-regulated transporter proteins) were involved in the increased uptake of Ni. Efficiency can be further reduced by slow growth of the hyperaccumulator and insufficient biomass.

Suboptimal efficiency in cultivation and processing can have a significant impact on agronomy. Some of the potential problems in this regard are

- inability to provide an optimal area for cultivation,
- insufficient number of seeds or seedlings,
- lack of or too expensive labor,
- unfavorable weather conditions,
- the time required for cultivation,
- problems with transport,
- disadvantages of technological processes,
- geographical isolation of the plant for combustion, which affects not only the cost of transporting the biomass but also the transport and utilization of the energy produced in this process, and
- fluctuations in the price of the extracted element.

3.8 Conclusion

Agromining is a promising and sustainable alternative for metal extraction, especially from low-grade ores and contaminated soils. Through the use of hyperaccumulators—plants that can uptake and accumulate high levels of metals in their leaves—phytomining represents an innovative solution to both environmental challenges and resource recovery. The unique characteristics of these plants, such as their ability to thrive in metalliferous soils while accumulating significant amounts of valuable metals, emphasize their potential for environmentally friendly mining methods.

Considering that agromining involves the cultivation of plants (in this case, hyperaccumulator species), the efficiency of agromining can be increased mainly through a careful selection of species and area of cultivation, as well as by the application of various agrotechnical measures. This makes the process more economically competitive. Soil management and amendments are also crucial for improving the bioavailability of metals, which has a direct impact on the efficiency of agromining. By optimizing soil conditions through targeted fertilization and irrigation, operating costs can be reduced while increasing the metal extraction capacity of the plants. In addition, research into cost-efficient techniques for harvesting and processing the metals from the plants will help to make the process more financially sustainable.

The future of agromining depends on continued interdisciplinary collaboration between experts in agronomy, biotechnology, and environmental science. By combining expertise from these fields, we can increase the performance of hyperaccumulators, improve extraction methods, and optimize the economic feasibility of phytomining. As research progresses, the techniques and strategies discussed in this chapter will play a critical role in making agromining a valuable tool for sustainable mining and land restoration, contributing to a more circular and environmentally sustainable economy.

References

Álvarez-López, V., Prieto-Fernández, Á., Cabello-Conejo, M.I., Kidd, P.S., 2016. Organic amendments for improving biomass production and metal yield of Ni-hyperaccumulating plants. Science of the Total Environment 548–549, 370–379. Available at https://doi.org/10.1016/j.scitotenv.2015.12.147.

Alves, S., Nabais, C., Simões Gonçalves, Md.L., Correia dos Santos, M.M., 2011. Nickel speciation in the xylem sap of the hyperaccumulator Alyssum serpyllifolium ssp. lusitanicum growing on serpentine soils of northeast Portugal. Journal of Plant Physiology 168 (15), 1715–1722. Available at https://doi.org/10.1016/j.jplph.2011.04.004.

Anderson, C.W., Stewart, R.B., Moreno, F.N., Wreesmann, C.T., Gardea-Torresdey, J.L., Robinson, B.H., et al., 2003. Gold phytomining. Novel developments in a plant-based mining system. In: Proceedings of the Gold 2003 Conference: New Industrial Applications of Gold, vol. 2. World Gold Council and Canadian Institute of Mining, Metallurgy and Petroleum. pp. 35–45.

Babička, J., 1943. Gold in Lebewesen. Mikrochemie Vereinigt mit Mikrochimica Acta 31 (4), 201–253. Available at https://doi.org/10.1007/bf01417553.

Baker, A.J.M., 1981. Accumulators and excluders: strategies in the response of plants to heavy metals. Journal of Plant Nutrition 3 (1–4), 643–654. Available at https://doi.org/10.1080/01904168109362867.

Baker, A.J.M., Brooks, R.R., 1989. Terrestrial higher plants which hyperaccumulate metallic elements: a review of their distribution, ecology and phytochemistry. Biorecovery 1, 81–126.

Baker, A.J.M., Proctor, J., van Balgooy, M.M.J., Reeves, R.D., 1992. Hyperaccumulation of nickel by the ultramafic flora of Palawan, Republic of the Philippines. In: Proctor, J.,

Baker, A.J.M., Reeves, R.D. (Eds.), The Vegetation of Ultramafic (Serpentine) Soils. Intercept Ltd., Andover, UK, pp. 291–304.

Bani, A., Echevarria, G., Sulçe, S., Morel, J.L., 2015. Improving the agronomy of *Alyssum murale* for extensive phytomining: a five-year field study. International Journal of Phytoremediation 17 (2), 117–127. Available at https://doi.org/10.1080/15226514.2013.862204.

Bani, A., Echevarria, G., Zhang, X., Benizri, E., Laubie, B., Morel, J.L., et al., 2015. The effect of plant density in nickel-phytomining field experiments with *Alyssum murale* in Albania. Australian Journal of Botany 63 (2), 72–77. Available at https://doi.org/10.1071/BT14285.

Bani, A., Pavlova, D., Garrido-Rodríguez, B., Kidd, P.S., Konstantinou, M., Kyrkas, D., et al., 2021. Element case studies in the temperate/mediterranean regions of Europe: nickel. In: van der Ent, A., Baker, A.J.M., Echevarria, G., Simonnot, M.O., Morel, J.L. (Eds.), Agromining: Farming for Metals: Extracting Unconventional Resources Using Plants. Springer, Cham, pp. 341–363. Available at https://doi.org/10.1007/978-3-030-58904-2_16.

Barbaroux, R., Plasari, E., Mercier, G., Simonnot, M.O., Morel, J.L., Blais, J.F., 2012. A new process for nickel ammonium disulfate production from ash of the hyperaccumulating plant *Alyssum murale*. Science of the Total Environment 423, 111–119. Available at https://doi.org/10.1016/j.scitotenv.2012.01.063.

Barberon, M., Berthomieu, P., Clairotte, M., Shibagaki, N., Davidian, J.C., Gosti, F., 2008. Unequal functional redundancy between the two *Arabidopsis thaliana* high-affinity sulphate transporters SULTR1;1 and SULTR1;2. New Phytologist 180 (3), 608–619. Available at https://doi.org/10.1111/j.1469-8137.2008.02604.x.

Battle, T., Moats, M., Cocalia, V., Oosterhof, H., Alam, S., Allanore, A., et al., 2013. Laterites: still a frontier of nickel process development Ni–Co. Wiley, pp. 1–23. Available at http://doi.org/10.1002/9781118658826.ch1.

Bačeva Andonovska, K., Stafilov, T., Matevski, V., 2021. Accumulation abilities of endemic plant species from the vicinity of an As-Sb-Tl abandoned mine, Allchar, Kožuf Mountain. In: Balabanova, B., Stafilov, T. (Eds.), Contaminant Levels and Ecological Effects. Springer, Cham, pp. 375–402. Available at https://doi.org/10.1007/978-3-030-66135-9-13.

Broadhurst, C.L., Chaney, R.L., Angle, J.S., Erbe, E.F., Maugel, T.K., 2004. Nickel localization and response to increasing Ni soil levels in leaves of the Ni hyperaccumulator *Alyssum murale*. Plant and Soil 265 (1–2), 225–242. Available at https://doi.org/10.1007/s11104-005-0974-8.

Cao, D., Zhang, H., Wang, Y., Zheng, L., 2014. Accumulation and distribution characteristics of zinc and cadmium in the hyperaccumulator plant *Sedum plumbizincicola*. Bulletin of Environmental Contamination and Toxicology 93 (2), 171–176. Available at https://doi.org/10.1007/s00128-014-1284-8.

Cassayre, L., Hazotte, C., Laubie, B., Carvalho, W.M., Simonnot, M.-O., 2020. Combustion of nickel hyperaccumulator plants investigated by experimental and thermodynamic approaches. Chemical Engineering Research and Design 160, 162–174. Available at https://doi.org/10.1016/j.cherd.2020.06.003.

Chaney, R.L., 1983. Plant uptake of inorganic waste constituents. In: Parr, J.F., Marsh, P.B., Kla, J.M. (Eds.), Land Treatment of Hazardous Wastes. Noyes Data Corp., Park Ridge, NJ, pp. 50–76.

Chaney, R.L., Kaufman, D.D., Hornick, S.B., Parr, J.F., Sikora, L.J., Burge, W.D., et al., 1981. Review of information relevant to land treatment of hazardous wastes. Report to USEPA Solid and Hazardous Waste Research Division.

Chaney, R.L., Chen, K.Y., Li, Y.M., Angle, J.S., Baker, A.J.M., 2008. Effects of calcium on nickel tolerance and accumulation in *Alyssum* species and cabbage grown in nutrient solution. Plant and Soil 311 (1-2), 131–140. Available at https://doi.org/10.1007/s11104-008-9664-7.

Chaney, R.L., Baker, A.J.M., Morel, J.L., 2021. The long road to developing agromining/phytomining. In: van der Ent, A., Baker, A.J.M., Echevarria, G., Simonnot, M.O., Morel, J.L. (Eds.), Agromining: Farming for Metals: Extracting Unconventional Resources Using Plants. Springer, Cham, pp. 1–22. Available at https://doi.org/10.1007/978-3-030-58904-2_16.

Chardot, V., Massoura, S.T., Echevarria, G., Reeves, R.D., Morel, J.L., 2005. Phytoextraction potential of the nickel hyperaccumulators *Leptoplax emarginata* and *Bornmuellera tymphaea*. International Journal of Phytoremediation 7 (4), 323–335. Available at https://doi.org/10.1080/16226510500327186.

Charlier, J.B., Polese, C., Nouet, C., Carnol, M., Bosman, B., Krämer, U., et al., 2015. Zinc triggers a complex transcriptional and post-transcriptional regulation of the metal homeostasis gene FRD3 in *Arabidopsis* relatives. Journal of Experimental Botany 66 (13), 3865–3878. Available at https://doi.org/10.1093/jxb/erv188, http://jxb.oxfordjournals.org/.

Cid, C.V., Pignata, M.L., Rodriguez, J.H., 2020. Effects of co-cropping on soybean growth and stress response in lead-polluted soils. Chemosphere 246, 125833. Available at https://doi.org/10.1016/j.chemosphere.2020.125833.

Cid, C.V., Rodriguez, J.H., Salazar, M.J., Blanco, A., Pignata, M.L., 2016. Effects of co-cropping *Bidens pilosa* (L.) and *Tagetes minuta* (L.) on bioaccumulation of Pb in *Lactuca sativa* (L.) growing in polluted agricultural soils. International Journal of Phytoremediation 18 (9), 908–917. Available at https://doi.org/10.1080/15226514.2016.1156636.

Clauss, M.J., Koch, M.A., 2006. Poorly known relatives of *Arabidopsis thaliana*. Trends in Plant Science 11 (9), 449–459. Available at https://doi.org/10.1016/j.tplants.2006.07.005.

Corzo Remigio, A., Pošćić, F., Nkrumah, P.N., Edraki, M., Spiers, K.M., Brueckner, D., et al., 2022. Comprehensive insights in thallium ecophysiology in the hyperaccumulator *Biscutella laevigata*. Science of the Total Environment 838 (Pt 2), 155899. Available at https://doi.org/10.1016/j.scitotenv.2022.155899.

Cui, L.Q., Wu, L.H., Li, N., Li, S.L., Li, L.Q., Pan, G.X., et al., 2009. Effects of soil moisture on growth and uptake of heavy metals of *Sedum plumbizincicola*. Soils 41, 572–576.

Datta, S., Chaudhury, K., Mukherjee, P.K., 2015. Hyperaccumulators from the serpentines of Andaman, India. Australian Journal of Botany 63 (3-4), 243–251. Available at https://doi.org/10.1071/BT14244, http://www.publish.csiro.au/nid/52/aid/49.htm.

Deng, D.M., Shu, W.S., Zhang, J., Zou, H.L., Lin, Z., Ye, Z.H., et al., 2007. Zinc and cadmium accumulation and tolerance in populations of *Sedum alfredii*. Environmental Pollution 147 (2), 381–386. Available at https://doi.org/10.1016/j.envpol.2006.05.024.

Deng, L., Li, Z., Wang, J., Liu, H., Li, N., Wu, L., et al., 2016. Long-term field phytoextraction of zinc/cadmium contaminated soil by *Sedum plumbizincicola* under different agronomic strategies. International Journal of Phytoremediation 18 (2), 134–140. Available at https://doi.org/10.1080/15226514.2015.1058328.

Drozdova, I., Alekseeva-Popova, N., Dorofeyev, V., Bech, J., Belyaeva, A., Roca, N., 2019. A comparative study of the accumulation of trace elements in Brassicaceae plant species with phytoremediation potential. Applied Geochemistry 108, 104377. Available at https://doi.org/10.1016/j.apgeochem.2019.104377.

Durand, A., Piutti, S., Rue, M., Morel, J.L., Echevarria, G., Benizri, E., 2016. Improving nickel phytoextraction by co-cropping hyperaccumulator plants inoculated by plant growth promoting rhizobacteria. Plant and Soil 399 (1–2), 179–192. Available at https://doi.org/10.1007/s11104-015-2691-2.

Echevarria, G., Morel, J.L., Fardeau, J.C., Leclerc-Cessac, E., 1998. Assessment of phytoavailability of nickel in soils. Journal of Environmental Quality 27 (5), 1064–1070. Available at https://doi.org/10.2134/jeq1998.00472425002700050011x.

Francesconi, K., Visoottiviseth, P., Sridokchan, W., Goessler, W., 2002. Arsenic species in an arsenic hyperaccumulating fern, *Pityrogramma calomelanos*: a potential phytoremediator of arsenic-contaminated soils. Science of the Total Environment 284 (1–3), 27–35. Available at https://doi.org/10.1016/s0048-9697(01)00854-3.

Galey, M.L., van der Ent, A., Iqbal, M.C.M., Rajakaruna, N., 2017. Ultramafic geoecology of South and Southeast Asia. Botanical Studies 58 (1). Available at https://doi.org/10.1186/s40529-017-0167-9.

Gérard, E., Echevarria, G., Sterckeman, T., Morel, J.L., 2000. Cadmium availability to three plant species varying in cadmium accumulation pattern. Journal of Environmental Quality 29 (4), 1117–1123. Available at https://doi.org/10.2134/jeq2000.00472425002900040012x.

Gonzaga, M.I.S., Gonzaga Santos, J.A., Ma, L.Q., 2006. Arsenic phytoextraction and hyperaccumulation by fern species. Scientia Agricola 63 (1), 90–101. Available at https://doi.org/10.1590/S0103-90162006000100015.

Goudard, L., Blaudez, D., Sirguey, C., Purwadi, I., Invernon, V., Rouhan, G., et al., 2024. Prospecting for rare earth element (hyper)accumulators in the Paris Herbarium using X-ray fluorescence spectroscopy reveals new distributional and taxon discoveries. Annals of Botany 133 (4), 573–584. Available at https://doi.org/10.1093/aob/mcae011.

Halimaa, P., Blande, D., Baltzi, E., Aarts, M.G.M., Granlund, L., Keinänen, M., et al., 2019. Transcriptional effects of cadmium on iron homeostasis differ in calamine accessio of *Noccaea caerulescens*. Plant Journal 97 (2), 306–320. Available at https://doi.org/10.1111/tpj.14121.

Hanikenne, M., Talke, I.N., Haydon, M.J., Lanz, C., Nolte, A., Motte, P., et al., 2008. Evolution of metal hyperaccumulation required cis-regulatory changes and triplication of HMA4. Nature 453 (7193), 391–395. Available at https://doi.org/10.1038/nature06877.

Hou, S., Wang, X., Shafi, M., Penttinen, P., Xu, W., Ma, J., et al., 2019. Remediation efficacy of *Sedum plumbizincicola* as affected by intercropping of landscape plants and oxalic acid in urban cadmium contaminated soil. Journal of Soils and Sediments 19 (10), 3512–3520. Available at https://doi.org/10.1007/s11368-019-02309-3.

Hu, P., Wang, Y., Przybyłowicz, W.J., Li, Z., Barnabas, A., Wu, L., Luo, Y., Mesjasz-Przybyłowicz, J., 2014. Elemental distribution by cryo-micro-PIXE in the zinc and cadmium hyperaccumulator *Sedum plumbizincicola* grown naturally. Plant and Soil 388 (1-2), 267–282. Available at https://doi.org/10.1007/s11104-014-2321-4.

Jaffré, T., Brooks, R.R., Lee, J., Reeves, R.D., 1976. *Sebertia acuminata*: a hyperaccumulator of nickel from New Caledonia. Science (New York, N.Y.) 193 (4253), 579–580. Available at https://doi.org/10.1126/science.193.4253.579.

Jaffré, T., Pillon, Y., Thomine, S., Merlot, S., 2013. The metal hyperaccumulators from New Caledonia can broaden our understanding of nickel accumulation in plants. Frontiers in Plant Science 4. Available at https://doi.org/10.3389/fpls.2013.00279.

Jakovljević, K., Bačeva Andonovska, K., Salinitro, M., Mišljenović, T., van der Ent, A., 2025. Biogeochemical survey of the Allchar (North Macedonia) arsenic-thallium ore body: a focus on hyperaccumulator plants. Plant and Soil 513 (1). Available at https://doi.org/10.1007/s11104-025-07252-6.

Jakovljević, K., Bani, A., Pavlova, D., Konstantinou, M., Dimitrakopoulos, P.G., Kyrkas, D., et al., 2022. Hyperaccumulator plant discoveries in the Balkans: accumulation, distribution, and practical applications. Botanica Serbica 46 (2), 161–178. Available at https://doi.org/10.2298/BOTSERB2202161J.

Jakovljević, K., Mišljenović, T., Bačeva Andonovska, K., Echevarria, G., Charrois, L., van der Ent, A., 2024. Living at the edge of life: metallophytes from the most toxic arsenic-thallium tailings in the world (Allchar, North Macedonia). Plant and Soil 497 (1–2), 413–428. Available at https://doi.org/10.1007/s11104-023-06404-w.

Jakovljević, K., Mišljenović, T., van der Ent, A., Baker, A.J.M., Andrejić, G., Tomović, G., et al., 2023. Zinc (hyper)accumulation in *Cardamine waldsteinii* from discovery in the herbarium to validation in the field. Plant Biosystems: An International Journal Dealing with all Aspects of Plant Biology 157 (4), 851–857. Available at https://doi.org/10.1080/11263504.2023.2204318.

Jakovljević, K., Mišljenović, T., van der Ent, A., Baker, A.J.M., Invernón, V.R., Echevarria, G., 2024. Mining the herbarium for hyperaccumulators: discoveries of nickel and zinc (hyper)accumulation in the genus *Noccaea* (Brassicaceae) through X-ray fluorescence herbarium scanning. Ecological Research 39 (4), 450–459. Available at https://doi.org/10.1111/1440-1703.12448.

Jankowski, K.J., Budzyński, W.S., Kijewski, Ł., Zając, T., 2015. Biomass quality of brassica oilseed crops in response to sulfur fertilization. Agronomy Journal 107 (4), 1377–1391. Available at https://doi.org/10.2134/agronj14.0386.

Jiang, C.A., Wu, Q.T., Goudon, R., Echevarria, G., Morel, J.L., 2015. Biomass and metal yield of co-cropped *Alyssum murale* and *Lupinus albus*. Australian Journal of Botany 63 (2), 159–166. Available at https://doi.org/10.1071/BT14261.

Jiang, Ca, Wu, Q.T., Sterckeman, T., Schwartz, C., Sirguey, C., Ouvrard, S., et al., 2010. Co-planting can phytoextract similar amounts of cadmium and zinc to mono-cropping from contaminated soils. Ecological Engineering 36 (4), 391–395. Available at https://doi.org/10.1016/j.ecoleng.2009.11.005.

Kazakou, E., Adamidis, G.C., Baker, A.J.M., Reeves, R.D., Godino, M., Dimitrakopoulos, P.G., 2010. Species adaptation in serpentine soils in Lesbos Island (Greece): metal hyperaccumulation and tolerance. Plant and Soil 332 (1), 369–385. Available at https://doi.org/10.1007/s11104-010-0302-9.

Khan, A.M., Bakar, N.K.A., Bakar, A.F.A., Ashraf, M.A., 2017. Chemical speciation and bioavailability of rare earth elements (REEs) in the ecosystem: a review. Environmental Science and Pollution Research 24 (29), 22764–22789. Available at https://doi.org/10.1007/s11356-016-7427-1, http://www.springerlink.com/content/.

Kidd, P.S., Bani, A., Benizri, E., Gonnelli, C., Hazotte, C., Kisser, J., et al., 2018. Developing sustainable agromining systems in agricultural ultramafic soils for nickel recovery. Frontiers in Environmental Science 6, 369383. Available at https://doi.org/10.3389/fenvs.2018.00044.

Krämer, U., 2005. MTP1 mops up excess zinc in Arabidopsis cells. Trends in Plant Science 10 (7), 313–315. Available at https://doi.org/10.1016/j.tplants.2005.05.008.

Krämer, U., Talke, I.N., Hanikenne, M., 2007. Transition metal transport. FEBS Letters 581 (12), 2263–2272. Available at https://doi.org/10.1016/j.febslet.2007.04.010.

Kukier, U., Peters, C.A., Chaney, R.L., Angle, J.S., Roseberg, R.J., 2004. The effect of pH on metal accumulation in two *Alyssum* species. Journal of Environmental Quality 33 (6), 2090–2102. Available at https://doi.org/10.2134/jeq2004.2090.

Kyrkas, D., Mantzos, N., Beza, P., Papantzikos, V., Andreaki, C., Konstantourou, E.E., et al., 2024. Effects of different shading levels on the growth of *Bornmuellera emarginata*, a nickel hyperaccumulator for use in agromining. Ecological Research 39 (4), 460–470. Available at https://doi.org/10.1111/1440-1703.12456.

Kyrkas, D., Mantzos, N., Patakioutas, G., Echevarria, G., Filis, E., Dimitrakopoulos, P.G., et al., 2023. The seed germination properties of two hyperaccumulator plant species with the potential for Ni agromining. Botanica Serbica 47 (2), 227–233. Available at https://doi.org/10.2298/BOTSERB2302227K, http://botanicaserbica.bio.bg.ac.rs/.

Lai, Y., Wang, Q., Yang, L., Huang, B., 2006. Subcellular distribution of rare earth elements and characterization of their binding species in a newly discovered hyperaccumulator *Pronephrium simplex*. Talanta 70 (1), 26–31. Available at https://doi.org/10.1016/j.talanta.2005.12.062.

Laubie, B., Vaughan, J., Simonnot, M.O., 2021. Processing of hyperaccumulator plants to nickel products. In: van der Ent, A., Baker, A.J.M., Echevarria, G., Simonnot, M.O., Morel, J.L. (Eds.), Agromining: Farming for Metals: Extracting Unconventional Resources Using Plants. Springer, Cham, pp. 47–61. Available at https://doi.org/10.1007/978-3-030-58904-2_3.

Li, Y.M., Chaney, R.L., Brewer, E.P., Angle, J.S., Nelkin, J., 2003. Phytoextraction of nickel and cobalt by hyperaccumulator *Alyssum* species grown on nickel-contaminated soils. Environmental Science and Technology 37 (7), 1463–1468. Available at https://doi.org/10.1021/es0208963.

Li, Y.M., Chaney, R., Brewer, E., Roseberg, R., Angle, J.S., Baker, A., et al., 2003. Development of a technology for commercial phytoextraction of nickel: economic and technical considerations. Plant and Soil 249 (1), 107–115. Available at https://doi.org/10.1023/A:1022527330401.

Li, J.T., Gurajala, H.K., Wu, L.H., van der Ent, A., Qiu, R.L., Baker, A.J.M., et al., 2018. Hyperaccumulator plants from China: a synthesis of the current state of knowledge. Environmental Science and Technology 52 (21), 11980–11994. Available at https://doi.org/10.1021/acs.est.8b01060.

Li, J., Jia, Y., Dong, R., Huang, R., Liu, P., Li, X., et al., 2019. Advances in the mechanisms of plant tolerance to manganese toxicity. International Journal of Molecular Sciences 20 (20), 5096. Available at https://doi.org/10.3390/ijms20205096.

Li, N., Tang, M.D., Cui, L.Q., Wu, L.H., Luo, Y.M., 2010. Effects of light intensity on plant growth and zinc and cadmium uptake by *Sedum plumbizincicola*. Acta Pedologica Sinica 47, 370–373.

Li, N., Wu, L.H., Luo, Y.M., Tang, M.D., Tan, C.Y., Jiang, Y.G., et al., 2009. Effects of harvesting way of *Sedum plumbizincicola* on its zinc and cadmium uptake in a mixed heavy metal contaminated soil. Acta Pedologica Sinica 46, 725–728. Available at https://doi.org/10.11766/trxb200710150423.

Liang, J., Shohag, M.J.I., Yang, X., Tian, S., Zhang, Y., Feng, Y., et al., 2014. Role of sulfur assimilation pathway in cadmium hyperaccumulation by *Sedum alfredii* Hance. Ecotoxicology and Environmental Safety 100 (1), 159–165. Available at https://doi.org/10.1016/j.ecoenv.2013.10.026.

Liu, C., Liu, W.S., van der Ent, A., Morel, J.L., Zheng, H.X., Wang, G.B., et al., 2021. Simultaneous hyperaccumulation of rare earth elements, manganese and aluminum in Phytolacca americana in response to soil properties. Chemosphere 282, 131096. Available at https://doi.org/10.1016/j.chemosphere.2021.131096.

Ly, S.N., Echevarria, G., Aarts, M.G., Ouvrard, S., van der Ent, A., 2025. Physiological responses of the nickel hyperaccumulator *Bornmuellera emarginata* under varying nickel dose levels and pH in hydroponics. Plant and Soil 507 (1), 939–951. Available at https://doi.org/10.1007/s11104-024-06777-6.

López-Millán, A.F., Ellis, D.R., Grusak, M.A., 2005. Effect of zinc and manganese supply on the activities of superoxide dismutase and carbonic anhydrase in *Medicago truncatula* wild type and raz mutant plants. Plant Science 168 (4), 1015–1022. Available at https://doi.org/10.1016/j.plantsci.2004.11.018.

Ma, L.Q., Komar, K.M., Tu, C., Zhang, W., Cai, Y., Kennelley, E.D., 2001. A fern that hyperaccumulates arsenic. Nature 409 (6820), 579. Available at https://doi.org/10.1038/35054664.

Ma, Y., Prasad, M.N.V., Rajkumar, M., Freitas, H., 2011. Plant growth promoting rhizobacteria and endophytes accelerate phytoremediation of metalliferous soils. Biotechnology Advances 29 (2), 248–258. Available at https://doi.org/10.1016/j.biotechadv.2010.12.001.

Milner, M.J., Kochian, L.V., 2008. Investigating heavy-metal hyperaccumulation using *Thlaspi caerulescens* as a model system. Annals of Botany 102 (1), 3–13. Available at https://doi.org/10.1093/aob/mcn063.

Morel, J.L., 2013. Using plants to "micro-mine" metals. In: Mollier P (ed). https://www.researchgate.net/publication/273203016_Des_plantes_pour_l%27extraction_des_metaux.

Morrey, D.R., Balkwill, K., Balkwill, M.-J., 1989. Studies on serpentine flora: preliminary analyses of soils and vegetation associated with serpentinite rock formations in the south-eastern Transvaal. South African Journal of Botany 55 (2), 171–177. Available at https://doi.org/10.1016/s0254-6299(16)31203-0.

Morrissey, J., Baxter, I.R., Lee, J., Li, L., Lahner, B., Grotz, N., et al., 2009. The ferroportin metal efflux proteins function in iron and cobalt homeostasis in Arabidopsis. The Plant Cell 21 (10), 3326–3338. Available at https://doi.org/10.1105/tpc.109.069401, http://www.plantcell.org/cgi/reprint/21/10/3326.

Nkrumah, P.N., Baker, A.J.M., Chaney, R.L., Erskine, P.D., Echevarria, G., Morel, J.L., et al., 2016. Current status and challenges in developing nickel phytomining: an agronomic perspective. Plant and Soil 406 (1–2), 55–69. Available at https://doi.org/10.1007/s11104-016-2859-4.

Nkrumah, P.N., Chaney, R.L., Morel, J.L., 2021. Agronomy of 'Metal Crops' Used in Agromining. In: van der Ent, A., Baker, A.J.M., Echevarria, G., Simonnot, M.O., Morel, J.L. (Eds.), Agromining: Farming for Metals: Extracting Unconventional Resources Using Plants. Springer, Cham, pp. 23–46. Available at https://doi.org/10.1007/978-3-030-58904-2_2.

Nkrumah, P.N., Echevarria, G., Erskine, P.D., Chaney, R.L., Sumail, S., van der Ent, A., 2019. Growth effects in tropical nickel-agromining 'metal crops' in response to nutrient dosing. Journal of Plant Nutrition and Soil Science 182 (5), 715–728. Available at https://doi.org/10.1002/jpln.201800468.

Nkrumah, P.N., Navarrete Gutiérrez, D.M., Tisserand, R., van der Ent, A., Echevarria, G., Pollard, J.A., et al., 2021. Element case studies: nickel (tropical regions). In: van der Ent, A., Baker, A.J.M., Echevarria, G., Simonnot, M.O., Morel, J.L. (Eds.), Agromining: Farming for Metals: Extracting Unconventional Resources Using Plants. Springer, Cham, pp. 365–383. Available at https://doi.org/10.1007/978-3-030-58904-2_17.

Nkrumah, P.N., Tisserand, R., Chaney, R.L., Baker, A.J.M., Morel, J.L., Goudon, R., et al., 2019. The first tropical 'metal farm': Some perspectives from field and pot experiments. Journal of Geochemical Exploration 198, 114–122. Available at https://doi.org/10.1016/j.gexplo.2018.12.003.

Pardo, T., Rodríguez-Garrido, B., Saad, R.F., Soto-Vázquez, J.L., Loureiro-Viñas, M., Prieto-Fernández, Á., et al., 2018. Assessing the agromining potential of Mediterranean nickel-hyperaccumulating plant species at field-scale in ultramafic soils under humid-temperate climate. Science of the Total Environment 630, 275–286. Available at https://doi.org/10.1016/j.scitotenv.2018.02.229.

Paçarizi, M., Krasniqi, E., Zogaj, M., Ukaj, S., 2020. Hyper-accumulation of nickel by *Bornmuellera dieckii* Degen as an endemic plant species in Kosovo. In: Stanojević, J., Kerolli-Mustafa, M. (eds). International Scientific Conference. Theory to Practice as a Cognitive, Educational and Social Challenge. pp. 277–284.

Peng, J.S., Wang, Y.J., Ding, G., Ma, H.L., Zhang, Y.J., Gong, J.M., 2017. A pivotal role of cell wall in cadmium accumulation in the Crassulaceae hyperaccumulator *Sedum plumbizincicola*. Molecular Plant 10 (5), 771–774. Available at https://doi.org/10.1016/j.molp.2016.12.007.

Pilon-Smits, E., 2005. Phytoremediation. Annual Review of Plant Biology 56 (1), 15–39. Available at https://doi.org/10.1146/annurev.arplant.56.032604.144214.

Purwadi, I., Erskine, P.D., Hutahaean, B.P., Wijaya, T.R., Nurtjahya, E., van der Ent, A., 2025. Rare earth elements (REEs) in soils and plants of Bangka Island (Indonesia) focussing on (hyper)accumulation. Plant and Soil 507 (1), 417–431. Available at https://doi.org/10.1007/s11104-024-06735-2.

Quimado, M.O., Fernando, E.S., Trinidad, L.C., Doronila, A., 2015. Nickel-hyperaccumulating species of *Phyllanthus* (Phyllanthaceae) from the Philippines. Journal of Botany 63 (2), 103–110. Available at https://doi.org/10.1071/BT14284.

Reeves, R.D., Baker, A.J.M., Jaffré, T., Erskine, P.D., Echevarria, G., van der Ent, A., 2018. A global database for plants that hyperaccumulate metal and metalloid trace elements. New Phytologist 218 (2), 407–411. Available at https://doi.org/10.1111/nph.14907.

Reeves, R.D., Baker, A.J.M., Kelepertsis, A., 1997. The distribution and biogeochemistry of some serpentine plants of Greece. In: Jaffré, T., Reeves, R.D., Becquer, T. (eds.). Écologie des milieux sur roches ultramafi ques et sur sols métallifères, ORSTOM, Nouméa, Documents Scientifiques et Techniques No. III/2. pp 205−207.

Reeves, R.D., Brooks, R.R., 1983. European species of *Thlaspi* L. (Cruciferae) as indicators of nickel and zinc. Journal of Geochemical Exploration 18 (3), 275–283. Available at https://doi.org/10.1016/0375-6742(83)90073-0.

Reeves, R.D., Brooks, R.R., Macfarlane, R.M., 1981. Nickel uptake by Californian *Streptanthus* and *Caulanthus* with particular reference to the hyperaccumulator *S. polygaloides* gray (Brassicaceae). American Journal of Botany 68 (5), 708–712. Available at https://doi.org/10.1002/j.1537-2197.1981.tb12403.x.

Reeves, R.D., Brooks, R.R., Robert Press, J., 1980. Nickel accumulation by species of *Peltaria* Jacq. (Cruciferae). Taxon 29 (5–6), 629–633. Available at https://doi.org/10.2307/1220334.

Reeves, R.D., Schwartz, C., Morel, J.L., Edmondson, J., 2001. Distribution and metal-accumulating behavior of *Thlaspi caerulescens* and associated metallophytes in France. International Journal of Phytoremediation 3 (2), 145–172. Available at https://doi.org/10.1080/15226510108500054.

Reeves, R.D., van der Ent, A., Echevarria, G., Isnard, S., Baker, A.J.M., 2021. Global distribution and ecology of hyperaccumulator plants. In: van der Ent, A., Baker, A.J.M., Echevarria, G., Simonnot, M.O., Morel, J.L. (Eds.), Agromining: Farming for Metals: Extracting Unconventional Resources Using Plants. Springer, Cham, pp. 133–154. Available at https://doi.org/10.1007/978-3-030-58904-2_7.

Ren, J., Wu, L.H., Liu, H.Y., Luo, Y.M., 2013. Effect of amendments on phytoextraction efficiency and metal uptake of following vegetable in heavy metal contaminated soil. Soils 45, 1233–1238.

Robinson, B., Anderson, C., 2021. Agromining of thallium and noble metals. In: van der Ent, A., Baker, A.J.M., Echevarria, G., Simonnot, M.O., Morel, J.L. (Eds.), Agromining: Farming for Metals: Extracting Unconventional Resources Using Plants. Springer, Cham, pp. 415–423. Available at https://doi.org/10.1007/978-3-030-58904-2_20.

Robinson, B.H., Brooks, R.R., Clothier, B.E., 1999. Soil amendments affecting nickel and cobalt uptake by *Berkheya coddii*: potential use for phytomining and phytoremediation. Annals of Botany 84 (6), 689–694. Available at https://doi.org/10.1006/anbo.1999.0970.

Robinson, W.O., Whetstone, R., Scribner, B.F., 1938. The presence of rare earths in hickory leaves. Science (New York, N.Y.) 87 (2264), 470. Available at https://doi.org/10.1126/science.87.2264.470.

Rosenkranz, T., Hipfinger, C., Ridard, C., Puschenreiter, M., 2019. A nickel phytomining field trial using *Odontarrhena chalcidica* and *Noccaea goesingensis* on an Austrian serpentine soil. Journal of Environmental Management 242, 522–528. Available at https://doi.org/10.1016/j.jenvman.2019.04.073.

Rue, M., Paul, A.L.D., Echevarria, G., Van Der Ent, A., Simonnot, M.O., Morel, J.L., 2020. Uptake, translocation and accumulation of nickel and cobalt in: Berkheya coddii, a 'metal crop' from South Africa. Metallomics 12 (8), 1278–1289. Available at https://doi.org/10.1039/d0mt00099j.

Saad, R.F., Echevarria, G., Rodríguez-Garrido, B., Kidd, P., Benizri, E., 2021. A two-year field study of nickel-agromining using *Odontarrhena chalcidica* co-cropped with a legume on an ultramafic soil: temporal variation in plant biomass, nickel yields and taxonomic and bacterial functional diversity. Plant and Soil 461 (1–2), 471–488. Available at https://doi.org/10.1007/s11104-021-04834-y.

Shan, X., Wang, H., Zhang, S., Zhou, H., Zheng, Y., Yu, H., et al., 2003. Accumulation and uptake of light rare earth elements in a hyperaccumulator *Dicropteris dichotoma*. Plant Science 165 (6), 1343–1353. Available at https://doi.org/10.1016/s0168-9452(03)00361-3.

Shen, L.B., Wu, L.H., Han, X.R., Tan, W.N., Huang, Y.J., Luo, Y.M., et al., 2011. Effects of nutrient regulation and control on plant growth and Zn/Cd uptake by hyperaccumulator *Sedum plumbizincicola*. Soils 43, 221–225.

Singh, S.N., Verma, A., 2007. Phytoremediation of air pollutants: a review. Environmental Bioremediation Technologies. Springer Berlin Heidelberg, India, pp. 293–314. Available at http://www.springerlink.com/openurl.asp?genre=book&isbn=978-3-540-34790-3.

Spielmann, J., Ahmadi, H., Scheepers, M., Weber, M., Nitsche, S., Carnol, M., et al., 2020. The two copies of the zinc and cadmium ZIP6 transporter of *Arabidopsis halleri* have distinct effects on cadmium tolerance. Plant, Cell & Environment 43 (9), 2143–2157. Available at https://doi.org/10.1111/pce.13806.

Stein, R.J., Höreth, S., de Melo, J.R.F., Syllwasschy, L., Lee, G., Garbin, M.L., et al., 2017. Relationships between soil and leaf mineral composition are element-specific, environment-dependent and geographically structured in the emerging model *Arabidopsis halleri*. New Phytologist 213 (3), 1274–1286. Available at https://doi.org/10.1111/nph.14219.

Stevanovic, V., Tan, K., Iatrou, G., 2003. Distribution of the endemic Balkan flora on serpentine I.-obligate serpentine endemics. Plant Systematics and Evolution 242 (1–4), 149–170. Available at https://doi.org/10.1007/s00606-003-0044-8.

Tang, Y., Qiu, R., Zeng, X., Fang, X., Yu, F., Zhou, X., et al., 2009. Zn and Cd hyperaccumulating characteristics of *Picris divaricata* Vant. International Journal of Environment and Pollution 38 (1-2), 26–38. Available at https://doi.org/10.1504/IJEP.2009.026640.

Thomas, W.A., 1975. Accumulation of rare earths and circulation of cerium by mockernut hickory trees. Canadian Journal of Botany 53 (12), 1159–1165. Available at https://doi.org/10.1139/b75-139.

Tognacchini, A., Rosenkranz, T., van der Ent, A., Machinet, G.E., Echevarria, G., Puschenreiter, M., 2020. Nickel phytomining from industrial wastes: growing nickel hyperaccumulator plants on galvanic sludges. Journal of Environmental Management 254, 109798. Available at https://doi.org/10.1016/j.jenvman.2019.109798.

Vaculík, M., Mišljenović, T., Lukačová, Z., Jakovljević, K., Podar, D., Kováč, J., 2025. Zinc hyperaccumulation in plants: mechanisms and principles. In: Tripathi, D.K., Singh, V.P., Pandey, S., Sharma, S., Chauhan, D.K. (Eds.), Zinc in Plants. Current Knowledge and Recent Advances. Elsevier Academic Press, pp. 1–39. Available at https://doi.org/10.1016/B978-0-323-91314-0.00017-X.

Van De Mortel, J.E., Schat, H., Moerland, P.D., Van Themaat, E.V.L., van der Ent, S., Blankestijn, H., et al., 2008. Expression differences for genes involved in lignin, glutathione and sulphate metabolism in response to cadmium in *Arabidopsis thaliana* and the related Zn/Cd-hyperaccumulator *Thlaspi caerulescens*. Plant, Cell and Environment 31 (3), 301–324. Available at https://doi.org/10.1111/j.1365-3040.2007.01764.x.

van der Ent, A., Baker, A.J.M., Reeves, R.D., Chaney, R.L., Anderson, C.W.N., Meech, J.A., et al., 2015. Agromining: farming for metals in the future? Environmental Science and Technology 49 (8), 4773–4780. Available at https://doi.org/10.1021/es506031u.

van der Ent, A., Baker, A.J.M., Reeves, R.D., Pollard, A.J., Schat, H., 2013. Hyperaccumulators of metal and metalloid trace elements: facts and fiction. Plant and Soil 362 (1–2), 319–334. Available at https://doi.org/10.1007/s11104-012-1287-3.

van der Ent, A., Erskine, P., Sumail, S., 2015. Ecology of nickel hyperaccumulator plants fro ultramafic soils in Sabah (Malaysia). Chemoecology 25 (5), 243–259. Available at https://doi.org/10.1007/s00049-015-0192-7.

van der Ent, A., Mesjasz-Przybyłowicz, J., Przybyłowicz, W.J., Barnabas, A.D., de Jonge, M.D., Harris, H.H., 2022. Contrasting patterns of nickel distribution in the hyperaccumulators *Phyllanthus balgooyi* and *Phyllanthus rufuschaneyi* from Malaysian Borneo. Metallomics 14 (5), mfac020. Available at https://doi.org/10.1093/mtomcs/mfac020.

van der Ent, A., Pollard, J.A., Echevarria, G., Abubakari, F., Erskine, P.D., Baker, A.J.M., et al., 2021. Exceptional Uptake and Accumulation of Chemical Elements in Plants: Extending the Hyperaccumulation Paradigm. In: van der Ent, A., Baker, A.J.M., Echevarria, G., Simonnot, M.O., Morel, J.L. (Eds.), Agromining: Farming for Metals: Extracting unconventional resources using plants. Springer, Cham, pp. 99–131. Available at https://doi.org/10.1007/978-3-030-58904-2_6.

Visioli, G., Vamerali, T., Mattarozzi, M., Dramis, L., Sanangelantoni, A.M., 2015. Combined endophytic inoculants enhance nickel phytoextraction from serpentine soil in the hyperaccumulator *Noccaea caerulescens*. Frontiers in Plant Science 6, 156553. Available at https://doi.org/10.3389/fpls.2015.00638.

Wu, L., Hu, P., Li, Z., Xu, W., Zhou, T., Zhong, D., et al., 2021. cElement case studies: cadmium and zinc. In: van der Ent, A., Baker, A.J.M., Echevarria, G., Simonnot, M.O., Morel, J.L. (Eds.), Agromining: Farming for Metals: Extracting Unconventional Resources Using Plants. Springer, Cham, pp. 453–469. Available at https://doi.org/10.1007/978-3-030-58904-2_23.

Wu, Y., Ma, L., Liu, Q., Sikder, M.M., Vestergård, M., Zhou, K., et al., 2020. Pseudomonas fluorescens promote photosynthesis, carbon fixation and cadmium phytoremediation of hyperaccumulator *Sedum alfredii*. Science of the Total Environment 726, 138554. Available at https://doi.org/10.1016/j.scitotenv.2020.138554.

Wu, J., Zhao, F.J., Ghandilyan, A., Logoteta, B., Guzman, M.O., Schat, H., et al., 2009. Identification and functional analysis of two ZIP metal transporters of the hyperaccumulator *Thlaspi caerulescens*. Plant and Soil 325 (1), 79–95. Available at https://doi.org/10.1007/s11104-009-0151-6.

Xiao, H., Zhang, Z., Li, F., Chai, Z., 2003. Study on contents and distribution characteristics of REE in fern by NAA. He Jishu/Nuclear Techniques 26 (6), 420. Available at http://hjs.periodicals.net.cn/default.html.

Xiaomei, L., Qitang, W., Banks, M.K., 2005. Effect of simultaneous establishment of Sedum alfredii and Zea mays on heavy metal accumulation in plants. International Journal of Phytoremediation 7 (1), 43–53. Available at https://doi.org/10.1080/16226510590915800.

Yuan, M., Guo, M.N., Liu, W.S., Liu, C., van der Ent, A., Morel, J.L., et al., 2017. The accumulation and fractionation of Rare Earth Elements in hydroponically grown *Phytolacca americana* L. Plant and Soil 421 (1–2), 67–82. Available at https://doi.org/10.1007/s11104-017-3426-3.

Zhang, X., Houzelot, V., Bani, A., Morel, J.L., Echevarria, G., Simonnot, M.O., 2014. Selection and combustion of Ni-hyperaccumulators for the phytomining process.

International Journal of Phytoremediation 16 (10), 1058–1072. Available at https://doi.org/10.1080/15226514.2013.810585.
Zhang, Z., Yu, Q., Du, H., Ai, W., Yao, X., Mendoza-Cózatl, D.G., et al., 2016. Enhanced cadmium efflux and root-to-shoot translocation are conserved in the hyperaccumulator *Sedum alfredii* (Crassulaceae family). FEBS Letters 590 (12), 1757–1764. Available at https://doi.org/10.1002/1873-3468.12225.
Zhao, B., Shen, L.B., Cheng, M.M., Wang, S.F., Wu, L.H., Zhou, S.B., et al., 2011. Effects of intercropping *Sedum plumbizincicola* in wheat growth season under wheat-rice rotation on the crops growth and their heavy metals uptake from different soil types. Chinese Journal of Applied Ecology 22 (10), 2725–2731.
Zhao, H., Wang, L., Zhao, F.-J., Wu, L., Liu, A., Xu, W., 2019. SpHMA1 is a chloroplast cadmium exporter protecting photochemical reactions in the Cd hyperaccumulator *Sedum plumbizincicola*. Plant, Cell & Environment 42 (4), 1112–1124. Available at https://doi.org/10.1111/pce.13456.

CHAPTER FOUR

Metal transporter in hyperaccumulator plant

4.1 Introduction

Heavy metals can be defined as those metals and metalloids having an atomic number >20, atomic weights in the range of 63.5–200.6 g mol^{-1}, density >5 g cc^{-1} and possess metallic properties. Some heavy metals are micronutrients (Co, Cu, Fe, Mn, Ni, and Zn), that is, essentially required for growth and metabolism in trace amounts (10–15 ppm). They also serve as cofactors in enzymatic reactions, but are toxic beyond certain concentrations. While others (Cd, Pb, Hg, Cr, and Al) have no known role in any metabolic activity and are toxic even at concentrations as low as 0.1% (Raychaudhuri et al., 2021). Plants normally cannot survive in soils containing these heavy metals without undergoing severe damage. However, certain plant members of Asteraceae (viz., *Berkheya coddii*), Brassicaceae (viz., *Arabidopsis halleri*, *Noccaea caerulescens*, and *Alyssum discolour*), Caryophyllaceae (viz., *Minuartia verna*, *Polycarpaea synandra*, and *Silene vulgaris*), Crassulaceae (viz., *Sedum alfredii*), Euphorbiaceae (viz., *Euphorbia macroclada*), Fabaceae (viz., *Anthyllis vulneraria*), Poaceae (viz., *Phragmites australis*, *Zea mays*, and *Agrostis stolonifera*), Plumbaginaceae (viz., *Armeria elongata*), and Violaceae (viz., *Viola calaminaria*) families; some ferns (viz., *Pteris vittata*) and also lichens are known to have evolved special adaptive mechanisms to survive in such conditions. These plants are known as metallophytes. The employment of such plants to accumulate toxic elements can be considered a permanent remediation technique. Different parts of plants show differential gene regulation upon exposure to toxic elements. Besides being equipped with specialized transporter proteins for influx and efflux of toxic elements, plants also undergo secretion of chelates to form complexes as a part of tolerance or detoxification mechanism (Asare et al., 2023). Different factors influence mobility and immobilization of metals and rule dynamics of elements in soil. The effect of soil pH can be seen in acidic soils where metals are readily dissolved leading to enhanced bioavailability (Kicińska et al., 2022), whereas high organic matter limits

Metallophytes in Mineral Mining
DOI: https://doi.org/10.1016/B978-0-443-30110-0.00001-5

the mobility and bioavailability, since their negative charge has tendency to form complexes with cationic elements (Sarkar et al., 2021). Root exudates, namely, acetic, fumaric, oxalic, and malic acids are capable of enhancing solubilization of elements in soil and modulate bioavailability by forming complexes (Lapie et al., 2019).

Currently, 746 species acting as hyperaccumulators have been reported globally: 532 for Ni (e.g., *Alyssum murale* and *B. coddii*), 53 for Cu (e.g., *Aeolanthus biformifolius* and *Jatropha curcas*), 42 for Co (e.g., *Haumaniastrum robertii*), 42 for Mn (e.g., *Virotia neurophylla*), 41 for Se (e.g., *Astragalus bisulcatus*), 20 for Zn (e.g., *N. caerulescens*), 8 for Pb (e.g., *Noccaea rotondifolia*), 7 for Cd (e.g., *A. halleri*, *Brassica rapa*, *Solanum nigrum*, and *S. alfredii*), 5 for As (e.g., *P. vittata*), 2 for Tl (e.g., *Biscutella laevigata*), and 2 for rare earth elements (REEs), that is, La and Ce (e.g., *Dicranopteris linearis*) (Reeves et al., 2018) and plants like *Phyllanthus rufuschaneyi* (Ni), *Antidesma puncticulatum* (Mn), and *Rinorea bengalensis* (Co) being the recent ones (Fadzil et al., 2024). The criteria for being classified as a hyperaccumulator is different for different metals: an accumulation of 100 mg kg^{-1} of dry weight for Cd or As, 1000 mg kg^{-1} for Co, Cu, Cr, Al, Ni, or Pb, and 10,000 mg kg^{-1} for Mn, Fe, or Zn. Metal preferences are not alike in all hyperaccumulators. Even within a particular family like Brassicaceae, out of the 87 species, Ni is hyperaccumulated by 67, Zn by 15, and both the metals by 5 species (Prasad et al., 2003).

4.2 Evolution of metallophytes

Metallophytes are one of the most rare and distinguished plants known to have evolved on metal-rich substrates as a result of natural selection. Plant species available for colonization of such land face strong selective pressure by the metals present in soil. As a result, only those plants survive and reproduce who have developed metal tolerance to an extent. Over the years, this kind of selective pressure leads to the development of metal tolerance traits in plants or simply the development of metallophytes in the course of evolution (Whiting et al., 2002). The mechanisms of metal tolerance vary significantly in plants to the degree of adaptation or evolution. In some plants, metal tolerance is limited to simply internal adaptive measures like reduction in uptake of metals, and/or their detoxification in its tissues when grown in metal-rich substrate. However, in other cases, plants have evolved genetically and diverged to form remarkable plants, the metallophytes that show extraordinary mechanisms to survive in metalliferous soil.

Metal tolerance along with metal hyperaccumulation, to some extent, is genetically distinct, since metal excluders tolerate high metal concentrations without hyperaccumulating them. However, hyperaccumulator plant species are extremely tolerant to metals even in their nonmetallicolous populations. Evolution of hyperaccumulation has been an outcome natural selection and is suggestive of ecological relevance, unlike induced hyperaccumulation in artificially spiked medium. In contrast to a neutral genetic drift, evolution of hyperaccumulators is a result of positive, directional natural selection. A continuously disturbed homeostasis leads to the evolution of extreme but consistent phenotypes in nature, that is, hyperaccumulators (van der Ent et al., 2015). The molecular mechanisms known to participate in metal hyperaccumulation are acquired from principles governing metal homeostasis. Compared to related nonaccumulating species, genes participating in metal homeostasis show differential expression in hyperaccumulators as an outcome of either duplication of a gene and/or changes in its promoter activity. Some metalliferous plants are hyperaccumulators that amass extremely high concentration of metals in its tissues (>0.1% metal in dry mass) without showing symptoms of toxicity. Hyperaccumulator plants are able to accumulate and tolerate around 100 folds higher concentration of metals in comparison to non-hyperaccumulating species (Chaney and Baklanov, 2017).

Hyperaccumulators can be classified as obligate or facultative on the basis of their distribution and nature of accumulation. Obligate hyperaccumulators are generally localized in metalliferous soils and compulsorily show hyperaccumulation phenomenon, whereas facultative hyperaccumulators are more widespread and do not show hyperaccumulation when present in nonmetalliferous soils (Pollard et al., 2014).

Hyperaccumulators are one of the most intriguing plant communities ecologically. Although the tolerance present in metalliferous plants is determined genetically, crossing experiments have demonstrated that it can be modified adaptively. From biotechnology point of view, hyperaccumulators are exceptional models for understanding mechanisms of metal uptake, transportation, and sequestration. They can be investigated to understand evolutionary mechanisms and adaptability in extreme environments and exploited for phytoremediation and biofortification purposes (Clemens, 2017). Furthermore, hyperaccumulators apply metal accumulation as a strategy against herbivory, allelopathy, and biotic interactions (Leitenmaier and Küpper, 2013).

Hyperaccumulators have evolved different mechanisms to cope with accumulated metals in comparison to non-hyperaccumulator plants. The principal mechanism of hyperaccumulation relies on elevated active metal transport mediated via enhanced expression of specified transport proteins in different plant tissues. Metal-specific ways of complexation and compartmentalization exist in both hyperaccumulator and non-hyperaccumulator plants. For example, hyperaccumulated metals are bound to weak ligands, whereas stronger ligands are only employed during transient binding for vacuolar transport and sequestration in hyperaccumulators. Metals that are not hyperaccumulated are treated similar to non-hyperaccumulators, that is, bound to stronger ligands for complexation, which is the main strategy for detoxification in non-hyperaccumulator plants. Both thiol (e.g., glutathione, phytochelatins, and metallothioneins) and non-thiol (e.g., histidine and nicotianamine) ligands are employed in the process of metal detoxification.

The first pivotal step in hyperaccumulating plants is enhanced metal uptake and high translocation from root to shoot along with lesser sequestration into root vacuoles. This translates to bioconcentration factor >1, that is, ratio of concentration of metal in shoots to soil. Furthermore, loading, transportation, and distribution of heavy metals by xylem to final sequestration sites, that is, vacuoles of large epidermal cells are also crucial steps in metal hyperaccumulation. Hyperaccumulators are well equipped to store toxic metals efficiently without compromising any chances to hampering photosynthetic machinery. Since vacuoles are sites for enzymes such as phosphatases, proteinases, and lipases, which are not targets of heavy metal toxicity, they provide excellent site for heavy metal sequestration. In order to keep metal concentration of mesophyll cells as low as possible, heavy metals are particularly sequestered in vacuoles of large epidermal cells that do not contain chloroplasts so that photosynthesis is not inhibited at any stage due to heavy metal transport.

However, there are a few exceptions to this. For instance, for *A. halleri*, a Zn hyperaccumulator that also accumulates Cd to a limited extent, majority of the Cd is sequestered in mesophyll cells, also presenting a threat for photosynthetic inhibition induced by Cd (Küpper et al., 2007). Another example is *S. alfredii*, a Cd/Zn hyperaccumulator, also accumulates Cd in mesophyll cells. However, being a Crassulaceae, it has thick succulent leaves and possesses atypical large vacuoles in mesophyll.

This feature enables them to store Cd without harm to mesophyll cells in comparison to other plants. Other exceptions include plants such as *Gossia bidwillii* (Myrtaceae), *V. neurophylla*, *Macadamia integrifolia*, and Macadamia *tetraphylla* (Proteaceae), where Mn is primarily sequestered in palisade layers. The translocation and sequestration of Mn in photosynthetically active cells can be attributed to its demand in OEC (oxygen-evolving-complex) active center (Fernando et al., 2006a,b).

These features of hyperaccumulators have also been validated via genetic studies. Genes engaged in heavy metal transport are overexpressed up to 200 times in hyperaccumulators in comparison to nonaccumulator plants. Hyperaccumulators are characterized by bioaccumulation coefficient (ratio of concentration of metal accumulated in body to that of substrate) much higher than 1. To achieve this, the metals need to be taken up by roots, transported, and pumped into their sequestration sites against concentration gradient involving an energy-consuming system, that is, active transport system. Heavy metal concentration may reach several hundred millimoles per liter in vacuoles of epidermal cells, which also suggests the requirement of active pumping to sequestration sites (Küpper et al., 2009). Nonessential trace elements are absorbed by roots of plants via transporters meant for essential elements due to similarity in valence states along with ionic diameters. For example, As resembles P, Cd resembles Cu/Fe/Mn/Zn, and Se resembles S (Alford et al., 2010). Transporters designed for these essential elements will inevitably transport nonessential elements due to chemical similarity between them.

4.3 Metal transporters in plants

Plants are provided with specific kinds of transporter for metal uptake, translocation, and maintenance of cytoplasmic ion balance. Several heavy metal transporter protein families have been recognized till date Zrt/Irt-like protein (ZIP), natural resistance-associated macrophage proteins (Nramp), cation diffusion facilitator (CDF)/metal tolerance protein (MTP), and heavy metal atpase (HMA), to name a few. Transcriptomic research has validated that upregulated expression of >30 genes result in metal accumulation in hyperaccumulator *A. halleri* when compared to nonaccumulator Arabidopsis *thaliana* (Skuza et al., 2022). A physiological model of metal uptake and transportation process in plants is demonstrated in Fig. 4.1

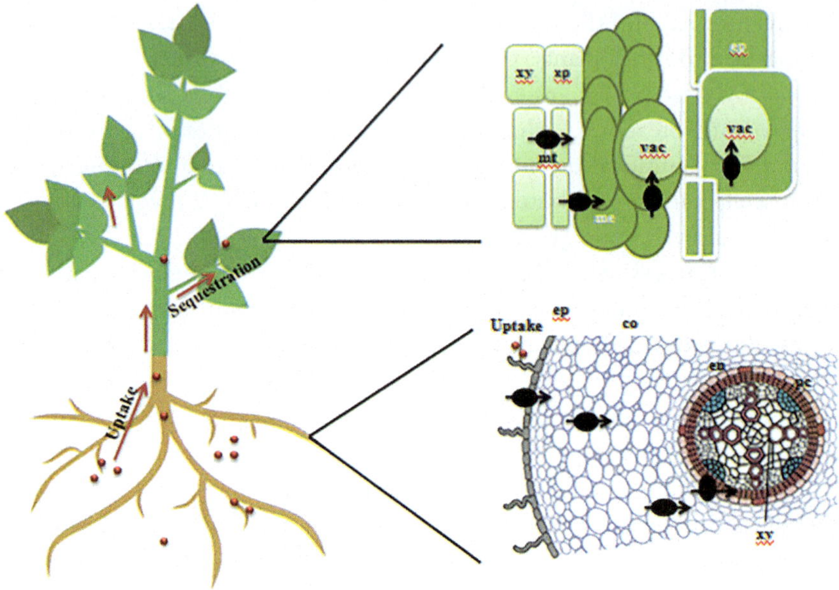

Figure 4.1 Physiological model of metal uptake and transportation process in plants (mt: metal transporter; ep: epidermis; co: cortex; me: mesophyll; vac: vacuole; en: endodermis; pc: pericycle; xy: xylem; xp: xylem parenchyma).

Recent studies on heavy metal transporter families in model plants have been enumerated in Table 4.1.

4.3.1 Zrt/Irt-like protein

The ZIP family is composed of ubiquitously expressed transporters critically associated with maintenance of systemic and cellular homeostasis of divalent metal ions. *ZIP* and associated transcription genes are involved in uptake, transport, detoxification, and homeostasis of Cd, Cu, Co, Ni, Fe, Mn, and especially Zn. *ZIP* genes are operative in cortex, endodermis, and pericycle and are also responsible for Zn^{2+} transport of cortex cells by diffusion. Zn^{2+} transport is hindered by apoplastic barrier and thus the transport to endodermis is mediated by ZIPs. Various transporters ZIP, HMA, and Yellow stripe-like (YSL) mediate Zn loading into the xylem from where it reaches leaf cells. Zn^{2+} may enter leaf cells freely or in chelation with citrate, histidine, or malate and undergoes vacuolar sequestration (Tao and Lu, 2022). Iron-regulated transporters (IRTs), a subfamily, are chiefly involved in transport of Fe and have been shown to

Table 4.1 Structural and functional characterization of heavy metal transporter gene families in model plants (*Arabidopsis* and rice).

Gene family	Structure	Members and their characterization in model plants	Function	References
ZIP (ZRT, IRT-like protein)	ZIP transporters in plants are anticipated to possess 6–9 transmembrane domains (α-helices). Molecular weight ranges from 33.1 to 51.4 kDa. Protein sequence ranges from 322 to 478 amino acids.	In *Arabidopsis*, 15 genes, i.e., *ZIP1–12, IRT1, IRT2*, and *IRT3* have been reported. Most ZIP members in *Arabidopsis thaliana* transport Zn^{2+}; with AtZIP3 and AtZIP9 being the exceptions. AtIRT1, AtIRT2, AtZIP7, and AtZIP9 transport Fe^{2+}; AtZIP1, AtZIP2, AtZIP5, AtZIP6, and AtZIP9 transport Mn^{2+}. In Oryza *sativa*, OsIRT1–2, OsZIP3–5, OsZIP8 transport Zn^{2+} and Fe^{2+}.	Involved in absorption and transportation of metal cations, such as Zn, Fe, Mn, and Cd; mainly distribution of Zn in whole plant.	Vatansever et al. (2016), Milner et al. (2013), and Zhang et al. (2023)
Nramp (natural resistance-associated macrophage proteins)	All Nramp proteins are reported have 11–12 transmembrane helices.	*NRAMP* gene family was first identified in rice among plants for Cd transport. NRAMPs are highly conserved genes that encode membrane proteins essential for transport of micronutrients including Fe, Zn, and Mn and stress response.	Found in the membranes of plant cells, such as the vacuolar membrane and the plastid envelope; transport a variety of transition metals such as Mn and Cd	Kanwal et al. (2024)

(*Continued*)

Table 4.1 (Continued)

Gene family	Structure	Members and their characterization in model plants	Function	References
		Their role has been well studied in *Arabidopsis* and rice. Six *NRAMP* genes have been recognized in *Arabidopsis* (*AtNRAMPs* 1–6).		
CDF/MTP (Cation diffusion facilitator/ metal tolerance protein)	Out of 12 MTP members in *Arabidopsis*, 6 AtMTPs (AtMTP1–5 and AtMTP12) belong to Zn-CDF cluster, while AtMTP6 and 7 are assigned to the Fe/Zn-CDF cluster. The remaining four AtMTPs (AtMTP8–11) are classified within the Mn-CDF cluster.	The first Mn-CDF to be functionally characterized in *A. thaliana* was *MTP11*. It is localized in pre-vacuolar or Golgi-like compartments, suggesting its role in conferring Mn tolerance. In rice, *MTP9* is mainly expressed in roots and is responsible for Mn translocation to stele. Ten MTPs have been identified in *O. sativa*.	*AtMTP1* confers tolerance to Zn by acting as a Zn/H^+ antiporter promoting vacuolar sequestration. *AtMTP3* increases Zn and Co tolerance by being involved in vacuolar transport; but is strongly induced during Fe deficiency so as to manage Zn uptake and maintain mineral balance.	Chu et al. (2017), Li et al.(2021), and Yilmaz et al. (2023))

HMA (Heavy metal atpase)	HMAs, also known as P_{1B}-ATPase, are a class of P-type ATPase superfamily and transport diverse kinds of ions across membranes coupled with ATP hydrolysis. HMAs contain 6–8 transmembrane helices. Phylogenetically, HMAs can be classified into 2 subgroups based on their specificity toward metal substrate: a Cu/Ag group and a Zn/Co/Cd/Pb group.	The genome of *A. thaliana* is found to encode 8 HMAs: *AtHMA1–4* belong to the Zn/Co/Cd/Pb subgroup. Nine *HMA* genes have been identified in rice and OsHMA1–3 belong to the Zn/Co/Cd/Pb subgroup.	HMAs are responsible for metal hyperaccumulation and conferring hypertolerance. AtHMA1 is localized in chloroplast and aids in excess Zn detoxification. AtHMA3 is localized in vacuolar membrane and leads to detoxification of Zn and Cd via sequestration in vacuoles. AtHMA2 and AtHMA4 are localized to the plasma membrane and aid in cellular efflux of Zn and Cd.	Takahashi et al. (2012)

(*Continued*)

Table 4.1 (Continued)

Gene family	Structure	Members and their characterization in model plants	Function	References
COPT/Ctr (Copper transporter)	COPT is characterized by the presence of 3 transmembrane domains, an extracellular methionine and/or histidine residue rich N-terminal region, an essential signature motif, an intracellular cysteine-rich C-terminal (except for COPT6).	Six members of the COPT family are encoded in *Arabidopsis* genome. AtCOPT1, 2, and 6 proteins reside on the plasma membrane and mediate Cu acquisition. In rice, COPT-type gene family consists of 7 members (*COPT1 –7*), out of which only *COPT1* and *5* have been functionally characterized.	*COPT1* mediates root Cu acquisition and is specifically expressed in the root tips. *COPT2* is particularly expressed in root differentiation zones only, not in elongation and meristematic zones. COPT5 protein functions in Cu mobilization from intracellular storage organelles. *COPT6* facilitates shoot Cu distribution and is expressed in aerial vascular tissues along with reproductive organs.	He et al. (2023) and Yuan et al. (2011)
MATE (Multidrug and toxic compound extrusion)	Members of the MATE family consist of 440–550 amino acids along with 12 transmembrane helices.	58 MATE gene orthologs have been identified in *Arabidopsis* and at least 53 in rice. AtDTX1 is the first multispecific MATE transporter identified in *A. thaliana*	AtDTX1 is not only involved in alkaloid, antibiotic, and toxic compound efflux but also detoxifies Cd^{2+}. OsMATE2 regulates the accumulation of As in rice and tobacco plants.	Wang et al. (2016) and Huang et al. (2021)

transport Cd in *Arabidopsis*; their overexpression enhances the uptake of Zn in shoots (Balafrej et al., 2020).

Root cells of Zn hyperaccumulators, for example, *A. halleri* and *N. caerulescens* show overexpression of *IRT3*. Gene transcriptions of different transporters (e.g., NRAMP3, MTP1 and 8, and HMA3 and 4) are also engaged in the vacuolar transport of Zn. Research has shown that Zn^{2+} transporter is overexpressed in hyperaccumulator plant species *N. caerulescens* when compared to Noccaea *arvense*, a nonaccumulator (Pence et al., 2000). The constitutive overexpression of metal transporter NcZNT1 in *A. thaliana* not only increased its tolerance but also accumulation upon excess Zn exposure when compared to wild-type plants (Lin et al., 2016). In the presence of Zn, the expressions of *ZIP19* and *ZIP23* have been found to be significantly upregulated in *A. halleri* and *N. caerulescens* root cells. There are high expressions of *ZIP19* and *ZIP23* in *A. thaliana* maintained under Zn-deficient culture conditions. Their upregulation in *A. halleri* and *N. caerulescens* could be related to low Zn concentration in the rhizosphere as a consequence of high activity of heavy metal ATPase 4 (HMA4) (involved in Zn depletion in roots by translocating it to aboveground parts) (Merlot et al., 2018).

4.3.2 Natural resistance-associated macrophage protein

They have been amply recognized in *Arabidopsis* and several other plants for their role in transportation of divalent ions Cd, Pb, and Zn (Bastow et al., 2018). They are naturally overexpressed in *N. caerulescens* and Oryza *sativa* and play a pivot role in Cd accumulation (Takahashi et al., 2011; Wei et al., 2009). While Nramp1, 3, and 4 regulate above-root accumulation in *A. thaliana*, Nramp6 mediates the transport of Cd from storage to toxic cellular compartments. In *N. caerulescens*, Nramp1 mediates Cd influx across plasma membrane of endodermis, hinting toward translocation from roots to shoots (Milner et al., 2014). In *Lolium multiflorum*, Pb, Co, and Ni transport has been shown to be mediated by NRAMP transporter family (Qiao et al., 2021).

AtNramp1, present on the plasma membrane, is a transporter with high affinity toward Mn. It is also involved in the transport of Fe and Cd in alliance with other proteins, namely, IRT1. AtNramp2, present in the trans-Golgi network, transports Mn to cytoplasm. Both AtNramp3 and AtNramp4 happen to be vacuolar metal transporters, maintaining the concentration of Fe and Mn between the vacuole and cytoplasm. AtNramp6 acts in intracellular migration and distribution of Cd, responsible for cadmium toxicity. Different Nramp proteins are in charge of transport of specific metal ions in rice. For example, OsNramp1 transports Fe, As, Mn, and Cd, whereas OsNramp3 is involved in the transport of Mn to young leaves under low-manganese conditions. OsNramp2 is responsible for the vacuolar transport of Fe similar to AtNramp3 and AtNramp4. While OsNramp4 specifically transports Al, OsNramp5 participates in the transportation of Fe, Mn, and Cd, whereas OsNramp6 transports Fe and Mn (Yang et al., 2023). *NRAMP3* gene has been shown to be highly expressed in *A. halleri* leaves (Weber et al., 2004).

4.3.3 Cation diffusion facilitators

These are known to play key role in metal homeostasis and imparting tolerance to plants. The CDF family has been categorized into three subfamilies, that is, Mn-CDF, Zn-CDF, and Zn/Fe-CDF, with regard to their substrate specificities. CDF family members in plants are named as MTPs and are responsible for the transport of Mn, Fe, and Zn mainly, besides being capable of transporting Cd, Co, Cu, and Ni. Several CDF transporters have been recognized (MTP, ZAT, and ZTP) that are homologous and responsible for the vacuolar storage of Zn and possibly Cd/Co. The first plant CDF transporter to be identified was AtMTP1, originally named ZAT (zinc transporter of *A. thaliana*), and was localized in tonoplast and mediated Zn detoxification along with its accumulation (Gu et al., 2021). MTP1 located on the vacuolar membrane has been shown to transport excess Co^{2+} into vacuoles in wheat plants (Wang et al., 2018). In *L. multiflorum, MTP1* was found to be upregulated in root cells upon exposure to Co^{2+}, Pb^{2+}, and Ni^{2+} (Qiao et al., 2021) for sequestration into vacuoles. *MTP8* is the most well functionally characterized MTP till date (Zhang et al., 2020) involved in Mn detoxification in *O. sativa* (Takemoto et al., 2017) and *A. thaliana* (Eroglu et al., 2016).

4.3.4 Heavy metal atpase

The HMA family are members of P-type ATPase superfamily and play a key part in the regulation of metal homeostasis in plants. Several *HMA* genes have been recognized in different plant species such as *Arabidopsis* and rice. *HMA3* and *HMA4* regulation was analyzed in Zn hyperaccumulators *A. halleri* and *N. caerulescens* at cellular as well as tissue level. *HMA3* and *HMA4* expressions were highest in the mesophyll of *A. halleri* but in the bundle sheath of vein in *N. caerulescens*, which explains the eventual storage sites for Zn in both the plants.

In *A. thaliana*, HMA3 mediates vacuolar sequestration of Cd, Co, Pb, and Zn in leaves (Morel et al., 2009). HMA3 is also responsible for the sequestration of Cd in *N. caerulescens* and *O. sativa* vacuoles (Ueno et al., 2011b). Since the localization of *HMA3* expression is unalike in both the plants, most Cd is sequestered in the shoots of *N. caerulescens* due to its overexpression there, whereas in *O. sativa*, most sequestration takes place in vacuoles of roots (Ueno et al., 2010, 2011a). Thus, an enhanced translocation from root to shoot as well as xylem unloading is seen in such cases.

The elevated expression of *HMA4* in hyperaccumulator plants' shoot besides root, being highest in the vascular bundles, also suggests that it may have a role in xylem unloading (Craciun et al., 2012). Its overexpression in nonaccumulator plants lead to severe heavy metal stress due to the lack of detoxifying mechanisms. In *N. caerulescens*, the constitutive upregulation of a Zn transporter is proposed to be generated, as the plant is constantly recognized being Zn-starved by the roots (Talke et al., 2006), and the upregulation of *HMA4* leads to an identical reaction in *A. halleri* (Hanikenne et al., 2008). RNA interference has also shown that HMA4 is responsible for the hyperaccumulation of Zn in *A. halleri*; its transfer enables Zn partitioning into xylem vessels in *A. thaliana* and upregulated specific genes characteristic for Zn hyperaccumulators (Hanikenne et al., 2008). AhHMA4 transporter present in the cell membrane removes cytoplasmic Cd and Zn outside the protoplast, while AtHMA3 is responsible for the transport of Zn^{2+}, Co^{2+}, and Pb^{2+} ions for vacuolar sequestration (Hasan et al., 2017).

4.3.5 ATP-binding cassette

The ATP-binding cassette (ABC) superfamily of transporters along with their subfamilies (ABCA–G) mediates heavy metal ion transport. They are also involved in phytochelatin (PC)mediated Cd detoxification by transporting chelated Cd to the vacuoles (Zhang et al., 2018). In rice, *ABCG36* enhances Cd tolerance by inhibiting its accumulation in root cells (Fu et al., 2019). *ABCG37* was found to be upregulated upon Cd exposure in Lolium *perenne* (Wang et al., 2020) and also by Co, Pb, and Ni, suggesting its involvement in translocation of metals by roots.

4.3.6 Multidrug-resistance related protein

Multidrug-resistance related protein (MRP) belongs to the ABCC subfamily and is involved in the transport of chelated Cd complexes into tonoplast. *MRP3* was found to be upregulated in genetically modified *Arabidopsis* and *Nicotiana* plants treated with As, Cd, Co, Ni, and Pb (Zientara et al., 2009) and also in *L. multiflorum* roots upon exposure to Co, Ni, and Pb (Qiao et al., 2021).

4.3.7 Yellow stripe-like

YSL proteins are responsible for the uptake and transport of Zn, Cd, and Fe ions. They play role in maintaining homeostasis of Zn and Fe in *Arabidopsis*, *O. sativa*, and *Hordeum vulgare* (Song et al., 2024). Presence of Cd in excess can stimulate the expression of *YSL*, and its overexpression has been shown to increases Cd translocation in *Arabidopsis* (Tao and Lu, 2022). Two classes of YSL transporters can be identified: One is typical for Poaceae plants, for example, *Z. mays*, *O. sativa, and H. vulgare*, and is involved in soil-to-root transport of $Fe^{2+/3+}$ phytosiderophore complexes. The other class plays a role in the transport and distribution of nicotianamine/phytosiderophore containing metal complexes intracellularly. These are present in *A. thaliana* (AtYSL1–4 and 6), *O. sativa* (OsYSL2, 6, 16, and 18), and *H. vulgare* (HvYSL2 and 5) (Skuza et al., 2022). The expression of *SnYSL3* (isolated from *S. nigrum*) occurs in the presence of excess Cd and increases over time, but the presence of Fe does not induce it. This indicates that the *SnYSL3* gene may not lead to the absorption of Fe from the soil. Although its overexpression occurs in vascular bundles, it has remarkable transport activity in other plant parts as well. The SnYSL3 protein is involved in the transport of metal–nicotianamine complex and regulates the stress response of plant caused due to excess Cd (Feng et al., 2017).

4.3.8 Nodulin 26-like intrinsic proteins

Nodulin 26-like intrinsic proteins (NIPs) are responsible for the transport of cationic elements As, Se, Si, and Sb. Various species of As are transported by plants through different transporters, for instance, NIP1, 2, 3, 5, and 7 are involved in the uptake of As(III) in *A. thaliana*, whereas the uptake of As(V) is done via phosphate transporters (PHTs) PHT1, 4, 5, 7, 8, and 9 (Xu et al., 2015).

In *P. vittata*, As (III) transport is facilitated by the tonoplast intrinsic protein (He et al., 2016).

4.3.9 Multidrug and toxic compound extrusion

Multidrug and toxic compound extrusion (MATE) transporters are constitutively overexpressed transmembrane proteins involved in the uptake, transfer, and sequestration of heavy metals in vacuoles. They are ubiquitously present in plants and have also been recognized in both archaea and bacteria. Besides being involved in the transit process of secondary metabolites and plant hormones, the members of MATE family are responsible for toxic metal efflux and aluminum tolerance (Huang et al., 2021).

4.3.10 High-affinity COPT/CTR (copper transporter)

High-affinity COPT/CTR have been recognized in both plants and mammals and that they play a key role in maintaining Cu homeostasis. COPT mediate the cellular entry of Cu ions across the plasma membrane, which is a passive process and is not coupled with an energy conversion process like ATP hydrolysis (He et al., 2023). The *Arabidopsis* genome is known to encode six members of the COPT family (COPT1–6), and COPT1 was the first member to be identified in plants. COPT2 has been shown to have potential for uptake and accumulation of Au in plants (Tiwari et al., 2017).

4.4 Conclusion and future prospects

The peculiar metal-trafficking abilities of hyperaccumulator plants render them likely candidates for application in remediation and phytomining projects. However, their applicability is limited due to slow growth cycle and survival issues of some plants under adverse climatic conditions or any other stress. This limitation can however be overcome by using engineered candidate plants, whereby metal transporter genes are overexpressed in native plants or microbes. The development of next-generation sequencing techniques has lead researchers to investigate the underlying mechanisms in metal hyperaccumulation and tolerance. Several studies have been conducted to transfer genes from hyperaccumulator plants to fast-growing resilient plant species. The investigation on the hyperaccumulation pattern of metals in unrelated plant families integrated with the collation of distinct accessions with disparate sequestration abilities within a species can possibly uncover principles behind the evolution of traits responsible for hyperaccumulation and hypertolerance. These studies may disclose the similarities and dissimilarities in mechanisms behind these traits, showing up evolutionary convergence and divergence. It would also unveil evolutionary "intermediates" (genotypes that are neither hyper-accumulators completely nor excluders) and will throw light on the order of the evolutionary events during the course of the adaptation of metal homeostasis network. The extent of pivotal gene expression responsible for hyperaccumulation could be utilized as a marker to anticipate metal sequestration abilities. Eventually, technologies, namely, CRISPR-CAS9 (Clustered regularly interspaced short palindromic repeats, CRISPR-associated protein 9) are currently being employed in plants for genome editing, that is, modifying the sequence of genes of interest (Doudna and Charpentier, 2014). The implementation of these technologies offers several benefits in comparison to conventional transformation technologies employed to create transgenic organisms; therefore, they are well received by the society and policymakers to design plants for phytoextraction. Based on the above technology, we can therefore, introduce point mutations in genetic sequences responsible for hyperaccumulation in order to improve their specificity and activity, or modify selectivity toward target metals (Pottier et al., 2015).

Genetic engineering offers the advantage of metal selectivity since transporters designed for hyperaccumulation are capable of identifying less abundant yet more toxic metals and can perform selective metal uptake. Another strategy under investigation is whether the downregulation or deletion of metal exporter genes from organelles such as Endoplasmic reticulum, golgi, and so on, can aid in higher metal retention and aid in metal compartmentalization. Moreover, analysis of different detoxification pathways such as glutathione biosynthesis and vacuolar/organellar metal compartmentalization can provide an insight into the mechanisms of hyperaccumulation that can be engineered to create transgenic candidates (Sun et al., 2019) that are more efficient than their counterparts. However, the complex multicellular organization and intricate gene clusters make plants difficult to engineer in comparison to single-celled counterparts, namely, bacteria and yeast. Although bacteria offer ease of manipulation and scalability, the absence of vacuoles makes yeast a better candidate for engineering at all levels, that is, from expressing specific transporter genes to complex metabolic pathways. Besides incorporating hyperactive metal membrane transporters in candidate plants, overexpression of vacuole transporters to compartmentalize metals from cytosol can also be a good strategy for designing transgenic hyperaccumulator species.

References

Alford, É.R., Pilon-Smits, E.A.H., Paschke, M.W., 2010. Metallophytes—a view from the rhizosphere. Plant and Soil 337 (1-2), 33–50. Available at 10.1007/s11104-010-0482-3.

Asare, M.O., Száková, J., Tlustoš, P., 2023. Mechanisms of As, Cd, Pb, and Zn hyperaccumulation by plants and their effects on soil microbiome in the rhizosphere. Frontiers in Environmental Science 11, 1157415. Available at 10.3389/fenvs.2023.1157415, https://www.journal.frontiersin.org/journal/environmental-science.

Balafrej, H., Bogusz, D., Triqui, Z.-E.A., Guedira, A., Bendaou, N., Smouni, A., et al., 2020. Zinc hyperaccumulation in plants: a review. Plants 9 (5), 562. Available at https://doi.org/10.3390/plants9050562.

Bastow, E.L., Garcia de la Torre, V.S., Maclean, A.E., Green, R.T., Merlot, S., Thomine, S., et al., 2018. Vacuolar iron stores gated by NRAMP3 and NRAMP4 are the primary source of iron in germinating seeds. Plant Physiology 177 (3), 1267–1276. Available at https://doi.org/10.1104/pp.18.00478, http://www.plantphysiol.org/content/177/3/1267.full.pdf.

Chaney, R.L., Baklanov, I.A., 2017. Phytoremediation and phytomining: status and promise. Advances in Botanical Research 83, 189–221. Available at http://www.elsevier.com/wps/find/bookdescription.cws_home/675801/description#description, https://doi.org/10.1016/bs.abr.2016.12.006, http://www.elsevier.com/wps/find/bookdescription.cws_home/675801/description#description.

Chu, H.H., Car, S., Socha, A.L., Hindt, M.N., Punshon, T., Guerinot, M.L., 2017. The Arabidopsis MTP8 transporter determines the localization of manganese and iron in seeds. Scientific Reports 7 (1), 11024. Available at https://doi.org/10.1038/s41598-017-11250-9, https://www.nature.com/srep/index.html.

Clemens, S., 2017. How metal hyperaccumulating plants can advance Zn biofortification. Plant and Soil 411 (1–2), 111–120. Available at https://doi.org/10.1007/s11104-016-2920-3, http://www.wkap.nl/journalhome.htm/0032-079X.

Craciun, A.R., Meyer, C.L., Chen, J., Roosens, N., De Groodt, R., Hilson, P., et al., 2012. Variation in HMA4 gene copy number and expression among Noccaea caerulescens populations presenting different levels of Cd tolerance and accumulation. Journal of Experimental Botany 63 (11), 4179–4189. Available at https://doi.org/10.1093/jxb/ers104.

Doudna, Charpentier, 2014. Genome editing. The new frontier of genome engineering with CRISPRCas9. Science (New York, N.Y.). 346.

van der Ent, A., Baker, A.J.M., Reeves, R.D., Pollard, A.J., Schat, H., 2015. Commentary: Toward a more physiologically and evolutionarily relevant definition of metal hyperaccumulation in plants. Frontiers in Plant Science 6, 144864. Available at http://journal.frontiersin.org/article/10.3389/fpls.2015.00554/pdf, https://doi.org/10.3389/fpls.2015.00554, http://journal.frontiersin.org/article/10.3389/fpls.2015.00554/pdf.

Eroglu, S., Meier, B., von Wirén, N., Peiter, E., 2016. The vacuolar manganese transporter MTP8 determines tolerance to iron deficiency-induced chlorosis in Arabidopsis. Plant Physiology 170 (2), 1030–1045. Available at https://doi.org/10.1104/pp.15.01194.

Fadzil, F.N.M., Mohamad, M.A.N., Repin, R., Harumain, Z.A.S., 2024. Metal uptake and tolerance in hyperaccumulator plants: advancing phytomining strategies. Rhizosphere 29, 100836. Available at https://www.sciencedirect.com/science/journal/24522198, https://doi.org/10.1016/j.rhisph.2023.100836, https://www.sciencedirect.com/science/journal/24522198.

Feng, S., Tan, J., Zhang, Y., Liang, S., Xiang, S., Wang, H., et al., 2017. Isolation and characterization of a novel cadmium-regulated Yellow Stripe-Like transporter (SnYSL3) in Solanum nigrum. Plant Cell Reports 36 (2), 281–296. Available at https://doi.org/10.1007/s00299-016-2079-7.

Fernando, D.R., Bakkaus, E.J., Perrier, N., Baker, A.J.M., Woodrow, I.E., Batianoff, G.N., et al., 2006a. Manganese accumulation in the leaf mesophyll of four tree species: a PIXE/EDAX localization study. New Phytologist 171 (4), 751–758. Available at https://doi.org/10.1111/j.1469-8137.2006.01783.x.

Fernando, D.R., Batianoff, G.N., Baker, A.J., Woodrow, I.E., 2006b. In vivo localization of manganese in the hyperaccumulator Gossia bidwillii (Benth.) N. Snow & Guymer (Myrtaceae) by cryo-SEM/EDAX. Plant, Cell and Environment 29 (5), 1012–1020. Available at https://doi.org/10.1111/j.1365-3040.2006.01498.x.

Fu, S., Lu, Y., Zhang, X., Yang, G., Chao, D., Wang, Z., et al., 2019. The ABC transporter ABCG36 is required for cadmium tolerance in rice. Journal of Experimental Botany 70 (20), 5909–5918. Available at https://doi.org/10.1093/jxb/erz335, http://jxb.oxfordjournals.org/.

Gu, D., Zhou, X., Ma, Y., Xu, E., Yu, Y., Liu, Y., et al., 2021. Expression of a Brassica napus metal transport protein (BnMTP3) in Arabidopsis thaliana confers tolerance to Zn and Mn. Plant Science 304, 110754. Available at https://doi.org/10.1016/j.plantsci.2020.110754.

Hanikenne, M., Talke, I.N., Haydon, M.J., Lanz, C., Nolte, A., Motte, P., et al., 2008. Evolution of metal hyperaccumulation required cis-regulatory changes and triplication of HMA4. Nature 453 (7193), 391–395. Available at https://doi.org/10.1038/nature06877, http://www.nature.com/nature/index.html.

Hasan, M.K., Cheng, Y., Kanwar, M.K., Chu, X.Y., Ahammed, G.J., Qi, Z.Y., 2017. Responses of plant proteins to heavy metal stress—a review. Frontiers in Plant Science 8, 282807. Available at http://journal.frontiersin.org/article/10.3389/fpls.2017.01492/full, https://doi.org/10.3389/fpls.2017.01492, http://journal.frontiersin.org/article/10.3389/fpls.2017.01492/full.

He, L., Ma, H., Song, W., Zhou, Z., Ma, C., Zhang, H., 2023. Arabidopsis COPT1 copper transporter uses a single histidine to regulate transport activity and protein stability. International Journal of Biological Macromolecules 241, 124404. Available at https://doi.org/10.1016/j.ijbiomac.2023.124404.

He, Z., Yan, H., Chen, Y., Shen, H., Xu, W., Zhang, H., et al., 2016. An aquaporin PvTIP4;1 from Pteris vittata may mediate arsenite uptake. New Phytologist 209 (2), 746–761. Available at https://doi.org/10.1111/nph.13637, http://www.wiley.com/bw/editors.asp?ref=0028-646X&site=1.

Huang, Y., He, G., Tian, W., Li, D., Meng, L., Wu, D., et al., 2021. Genome-wide identification of MATE gene family in potato (Solanum tuberosum L.) and expression analysis in heavy metal stress. Frontiers in Genetics 12, 650500. Available at https://doi.org/10.3389/fgene.2021.650500.

Kanwal, F., Riaz, A., Ali, S., Zhang, G., 2024. NRAMPs and manganese: magic keys to reduce cadmium toxicity and accumulation in plants. Science of The Total Environment 921, 171005. Available at https://doi.org/10.1016/j.scitotenv.2024.171005.

Kicińska, A., Pomykała, R., Izquierdo-Diaz, M., 2022. Changes in soil pH and mobility of heavy metals in contaminated soils. European Journal of Soil Science 73 (1), e13203. Available at http://onlinelibrary.wiley.com/journal/10.1111/(ISSN)1365-2389, https://doi.org/10.1111/ejss.13203, http://onlinelibrary.wiley.com/journal/10.1111/(ISSN)1365-2389.

Küpper, H., Götz, B., Mijovilovich, A., Küpper, F.C., Meyer-Klaucke, W., 2009. Complexation and toxicity of copper in higher plants. I. Characterization of copper accumulation, speciation, and toxicity in Crassula helmsii as a new copper accumulator. Plant Physiology 151 (2), 702–714. Available at https://doi.org/10.1104/pp.109.139717.

Küpper, H., Parameswaran, A., Leitenmaier, B., Trtílek, M., Šetlík, I., 2007. Cadmium-induced inhibition of photosynthesis and long-term acclimation to cadmium stress in the hyperaccumulator Thlaspi caerulescens. New Phytologist 175 (4), 655–674. Available at https://doi.org/10.1111/j.1469-8137.2007.02139.x.

Lapie, C., Leglize, P., Paris, C., Buisson, T., Sterckeman, T., 2019. Profiling of main metabolites in root exudates and mucilage collected from maize submitted to cadmium stress. Environmental Science and Pollution Research 26 (17), 17520–17534. Available at https://doi.org/10.1007/s11356-019-05168-0.
Leitenmaier, B., Küpper, H., 2013. Compartmentation and complexation of metals in hyperaccumulator plants. Frontiers in Plant Science 4, 55114. Available at https://doi.org/10.3389/fpls.2013.00374.
Lin, Y.-F., Hassan, Z., Talukdar, S., Schat, H., Mark, G.M., Aarts, I., 2016. Baxter, expression of the ZNT1 zinc transporter from the metal hyperaccumulator Noccaea caerulescens confers enhanced zinc and cadmium tolerance and accumulation to Arabidopsis thaliana. PLoS ONE 11 (3), e0149750. Available at https://doi.org/10.1371/journal.pone.0149750.
Li, J., Dong, R., Jia, Y., Huang, J., Zou, X., An, N., et al., 2021. Characterization of metal tolerance proteins and functional analysis of GmMTP8.1 involved in manganese tolerance in soybean. Frontiers in Plant Science 12, 683813. Available at https://www.frontiersin.org/journals/plant-science, https://doi.org/10.3389/fpls.2021.683813, https://www.frontiersin.org/journals/plant-science.
Merlot, S., de la Torre, V.S.G., Hanikenne, M., 2018. Physiology and molecular biology of trace element hyperaccumulation. Agromining: Farming for Metals. Springer: Berlin, Germany, pp. 93–116.
Milner, M.J., Mitani-Ueno, N., Yamaji, N., Yokosho, K., Craft, E., Fei, Z., et al., 2014. Root and shoot transcriptome analysis of two ecotypes of Noccaea caerulescens uncovers the role of NcNramp1 in Cd hyperaccumulation. Plant Journal 78 (3), 398–410. Available at https://doi.org/10.1111/tpj.12480, http://www.blackwellpublishing.com/journals/TPJ.
Milner, M.J., Seamon, J., Craft, E., Kochian, L.V., 2013. Transport properties of members of the ZIP family in plants and their role in Zn and Mn homeostasis. Journal of Experimental Botany 64 (1), 369–381. Available at https://doi.org/10.1093/jxb/ers315.
Morel, M., Crouzet, J., Gravot, A., Auroy, P., Leonhardt, N., Vavasseur, A., et al., 2009. AtHMA3, a P1B-ATPase allowing Cd/Zn/Co/Pb vacuolar storage in Arabidopsis. Plant Physiology 149 (2), 894–904. Available at https://doi.org/10.1104/pp.108.130294.
Pence, N.S., Larsen, P.B., Ebbs, S.D., Letham, D.L.D., Lasat, M.M., Garvin, D.F., et al., 2000. The molecular physiology of heavy metal transport in the Zn/Cd hyperaccumulator Thlaspi caerulescens. Proceedings of the National Academy of Sciences of the United States of America 97 (9), 4956–4960. Available at https://doi.org/10.1073/pnas.97.9.4956.
Pollard, A.J., Reeves, R.D., Baker, A.J.M., 2014. Facultative hyperaccumulation of heavy metals and metalloids. Plant Science 217–218, 8–17. Available at https://doi.org/10.1016/j.plantsci.2013.11.011.
Pottier, M., Oomen, R., Picco, C., Giraudat, J., Scholz-Starke, J., Richaud, P., et al., 2015. Identification of mutations allowing natural resistance associated macrophage proteins (NRAMP) to discriminate against cadmium. Plant Journal 83 (4), 625–637. Available at https://doi.org/10.1111/tpj.12914.
Prasad, V., Narasimha, M., Freitas, de, O., Maria, H., 2003. Metal hyperaccumulation in plants: biodiversity prospecting for phytoremediation technology. Electronic Journal of Biotechnology 6, 285–321.
Qiao, S., Tao, Y., Shan, Q., Wang, J., Chai, T., Gong, S., et al., 2021. Physiological and gene expression responses of six annual ryegrass cultivars to cobalt, lead, and nickel stresses. International Journal of Molecular Sciences 22 (24), 13583. Available at https://doi.org/10.3390/ijms222413583, https://www.mdpi.com/1422-0067/22/24/13583/pdf.
Raychaudhuri, S.S., Pramanick, P., Talukder, P., Basak, A., 2021. Polyamines, metallothioneins, and phytochelatins—Natural defense of plants to mitigate heavy

metals. Studies in Natural Products Chemistry 69, 227–261. Available at https://doi.org/10.1016/B978-0-12-819487-4.00006-9, http://www.elsevier.com/wps/find/bookdescription.cws_home/715507/description#description.
Reeves, R.D., Baker, A.J.M., Jaffré, T., Erskine, P.D., Echevarria, G., van der Ent, A., 2018. A global database for plants that hyperaccumulate metal and metalloid trace elements. New Phytologist 218 (2), 407–411. Available at http://onlinelibrary.wiley.com/journal/10.1111/(ISSN)1469-8137, https://doi.org/10.1111/nph.14907, http://onlinelibrary.wiley.com/journal/10.1111/(ISSN)1469-8137.
Sarkar, B., Mukhopadhyay, R., Ramanayaka, S., Bolan, N., Ok, Y.S., 2021. The role of soils in the disposition, sequestration and decontamination of environmental contaminants. Philosophical Transactions of the Royal Society B: Biological Sciences 376 (1834), 20200177. Available at https://doi.org/10.1098/rstb.2020.0177.
Skuza, L., Szućko-Kociuba, I., Filip, E., Bożek, I., 2022. Natural molecular mechanisms of plant hyperaccumulation and hypertolerance towards heavy metals. International Journal of Molecular Sciences 23, 9335. Available at https://doi.org/10.3390/ijms23169335.
Song, Z., Li, S., Li, Y., Zhou, X., Liu, X., Yang, W., et al., 2024. Identification and characterization of yellow stripe-like genes in maize suggest their roles in the uptake and transport of zinc and iron. BMC Plant Biology 24 (1), 3. Available at https://doi.org/10.1186/s12870-023-04691-0.
Sun, G.L., Reynolds, E.E., Belcher, A.M., 2019. Designing yeast as plant-like hyperaccumulators for heavy metals. Nature Communications 10 (1), 5080. Available at https://doi.org/10.1038/s41467-019-13093-6, http://www.nature.com/ncomms/index.html.
Takahashi, R., Bashir, K., Ishimaru, Y., Nishizawa, N.K., Nakanishi, H., 2012. The role of heavy-metal ATPases, HMAs, in zinc and cadmium transport in rice. Plant Signaling and Behavior 7 (12), 1605–1607. Available at https://doi.org/10.4161/psb.22454, http://www.landesbioscience.com/journals/psb/article/22454/?download_pdf=true.
Takahashi, R., Ishimaru, Y., Senoura, T., Shimo, H., Ishikawa, S., Arao, T., et al., 2011. The OsNRAMP1 iron transporter is involved in Cd accumulation in rice. Journal of Experimental Botany 62 (14), 4843–4850. Available at https://doi.org/10.1093/jxb/err136.
Takemoto, Y., Tsunemitsu, Y., Fujii-Kashino, M., Mitani-Ueno, N., Yamaji, N., Ma, J.F., et al., 2017. The tonoplast-localized transporter MTP8.2 contributes to manganese detoxification in the shoots and roots of oryza sativa L. Plant and Cell Physiology 58 (9), 1573–1582. Available at https://doi.org/10.1093/pcp/pcx082, http://pcp.oxfordjournals.org/.
Talke, I.N., Hanikenne, M., Krämer, U., 2006. Zinc-dependent global transcriptional control, transcriptional deregulation, and higher gene copy number for genes in metal homeostasis of the hyperaccumulator Arabidopsis halleri. Plant Physiology 142 (1), 148–167. Available at https://doi.org/10.1104/pp.105.076232.
Tao, J., Lu, L., 2022. Advances in genes-encoding transporters for cadmium uptake, translocation, and accumulation in plants. Toxics 10 (8), 411. Available at https://doi.org/10.3390/toxics10080411.
Tiwari, M., Venkatachalam, P., Penarrubia, L., Sahi, S.V., 2017. COPT2, a plasma membrane located copper transporter, is involved in the uptake of Au in Arabidopsis. Scientific Reports 7 (1), 11430. Available at https://doi.org/10.1038/s41598-017-11896-5, https://www.nature.com/srep/index.html.
Ueno, D., Koyama, E., Yamaji, N., Ma, J.F., 2011b. Physiological, genetic, and molecular characterization of a high-Cd-accumulating rice cultivar, Jarjan. Journal of Experimental Botany 62 (7), 2265–2272. Available at https://doi.org/10.1093/jxb/erq383.
Ueno, D., Millner, M.J., Yamaji, N., Yokosho, K., Koyama, E., Zambrono, M.C., et al., 2011a. Elevated expression of TcHMA3 plays a key role in the extreme cadmium

tolerance in a Cd-hyperaccumulating ecotype of Thlaspi caerulescens. The Plant Journal 66, 852–862. Available at https://doi.org/10.1111/j.1365-313X.2011.04548.x.

Ueno, D., Yamaji, N., Kono, I., Huang, C.F., Ando, T., Yano, M., et al., 2010. Gene limiting cadmium accumulation in rice. Proceedings of the National Academy of Sciences of the United States of America 107 (38), 16500–16505. Available at https://doi.org/10.1073/pnas.1005396107, http://www.pnas.org/content/107/38/16500.full.pdf+html.

Vatansever, R., Özyiğit, İ.İ., Filiz, E., 2016. Comparative and phylogenetic analysis of zinc transporter genes/proteins in plants. Turkish Journal of Biology 40 (3), 600–611. Available at https://doi.org/10.3906/biy-1501-91, http://journals.tubitak.gov.tr/biology/issues/biy-16-40-3/biy-40-3-9-1501-91.pdf.

Wang, L., Bei, X., Gao, J., Li, Y., Yan, Y., Hu, Y., 2016. The similar and different evolutionary trends of MATE family occurred between rice and Arabidopsis thaliana. BMC Plant Biology 16 (1), 207. Available at https://doi.org/10.1186/s12870-016-0895-0.

Wang, F.H., Qiao, K., Liang, S., Tian, S.Q., Tian, Y.B., Wang, H., et al., 2018. Triticum urartu MTP1: its ability to maintain Zn2+ and Co2+ homeostasis and metal selectivity determinants. Plant Cell Reports 37 (12), 1653–1666. Available at https://doi.org/10.1007/s00299-018-2336-z, https://www.link.springer.de/link/service/journals/00299/index.htm.

Wang, J., Zhao, J., Feng, S., Zhang, J., Gong, S., Qiao, K., et al., 2020. Comparison of cadmium uptake and transcriptional responses in roots reveal key transcripts from high and low-cadmium tolerance ryegrass cultivars. Ecotoxicology and Environmental Safety 203, 110961. Available at https://doi.org/10.1016/j.ecoenv.2020.110961, http://www.elsevier.com/inca/publications/store/6/2/2/8/1/9/index.htt.

Weber, M., Harada, E., Vess, C., Roepenack-Lahaye, E.V., Clemens, S., 2004. Comparative microarray analysis of Arabidopsis thaliana and Arabidopsis halleri roots identifies nicotianamine synthase, a ZIP transporter and other genes as potential metal hyperaccumulation factors. Plant Journal 37 (2), 269–281. Available at https://doi.org/10.1046/j.1365-313X.2003.01960.x, http://www.blackwellpublishing.com/journals/TPJ.

Wei, W., Chai, T., Zhang, Y., Han, L., Xu, J., Guan, Z., 2009. The Thlaspi caerulescens NRAMP homologue TcNRAMP3 is capable of divalent cation transport. Molecular Biotechnology 41 (1), 15–21. Available at https://doi.org/10.1007/s12033-008-9088-x.

Whiting, S.N., Reeves, R.D., Baker, A.J.M., 2002. Conserving biodiversity: mining, metallophytes and land reclamation. Mining Environmental Management 10 (2), 11–16.

Xu, W., Dai, W., Yan, H., Li, S., Shen, H., Chen, Y., et al., 2015. Arabidopsis NIP3;1 plays an important role in arsenic uptake and root-to-shoot translocation under arsenite stress conditions. Molecular Plant 8 (5), 722–733. Available at https://doi.org/10.1016/j.molp.2015.01.005.

Yang, Y., Zheng, J., Liang, Y., Wang, X., Li, K., Chen, L., et al., 2023. Natural resistance-associated macrophage protein (Nramp) family in foxtail millet (Setaria italica): characterization, expression analysis and relationship with metal content under Cd stress. Agronomy 13 (8), 2000.

Yilmaz, H., Özer, G., Baloch, F.S., Çiftçi, V., Chung, Y.S., Sun, H.J., 2023. Genome-wide identification and expression analysis of MTP (metal ion transport proteins) genes in the common bean. Plants 12 (18), 3218. Available at https://doi.org/10.3390/plants12183218, https://www.mdpi.com/journal/plants.

Yuan, M., Li, X., Xiao, J., Wang, S., 2011. Molecular and functional analyses of COPT/Ctr-type copper transporter-like gene family in rice. BMC Plant Biology 11 (1), 69. Available at https://doi.org/10.1186/1471-2229-11-69.

Zhang, X., Li, Q., Xu, W., Zhao, H., Guo, F., Wang, P., et al., 2020. Identification of MTP gene family in tea plant (Camellia sinensis L.) and characterization of CsMTP8.2

in manganese toxicity. Ecotoxicology and Environmental Safety 202, 110904. Available at https://doi.org/10.1016/j.ecoenv.2020.110904.

Zhang, J., Martinoia, E., Lee, Y., 2018. Vacuolar transporters for cadmium and arsenic in plants and their applications in phytoremediation and crop development. Plant and Cell Physiology 59 (7), 1317–1325. Available at https://doi.org/10.1093/pcp/pcy006, http://pcp.oxfordjournals.org/.

Zhang, Z., Wang, M., Zhang, X., Wang, W., He, X., Wang, R., et al., 2023. Genome-wide identification and expression analysis of the ZIP gene family in Quercus dentata. Current Plant Biology 35–36, 100291. Available at https://doi.org/10.1016/j.cpb.2023.100291.

Zientara, K., Wawrzyńska, A., Łukomska, J., López-Moya, J.R., Liszewska, F., Assunção, A.G.L., et al., 2009. Activity of the AtMRP3 promoter in transgenic Arabidopsis thaliana and Nicotiana tabacum plants is increased by cadmium, nickel, arsenic, cobalt and lead but not by zinc and iron. Journal of Biotechnology 139 (3), 258–263. Available at https://doi.org/10.1016/j.jbiotec.2008.12.001.

CHAPTER FIVE

Microbe-assisted phytomining

5.1 Introduction

Phytoremediation, as a latest research discipline was introduced in 1970s and subsequently evolved with the introduction and involvement of hyperaccumulator plants in 21st century. The progression of "phytoremediation" to "phytomining" to reclaim metals for revenue generation is afoot (Jiang et al., 2015). Various research studies have identified 746 hyperaccumulator plant species spanning across 52 families (e.g. Asteraceae, Brassicaceae, Caryophyllaceae, Cyperaceae, Cumouniceae, Fabaceae, Flacourtaceae, Lamiaceae, Phyllanthaceae, Poaceae, Violaceae, and Euphorbiaceae) and 130 genera, out of which Brassicaceae and Phyllanthaceae represent the greatest number of hyperaccumulator species, that is, 83 and 59, respectively (Salinitro et al., 2022; Reeves et al., 2017).

Phytomining involves cultivating those plants that are capable of taking up trace metals from metalliferous soil and translocating to shoots (>1%) that later serve as a bio-ore and can be harvested to recover those metals (Echevarria et al., 2015). The initiation of agro/phytomining project requires (1) identification/selection of hyperaccumulator species, preferably possessing high biomass; (2) evaluating its genetic diversity, followed by breeding improved strains for a greater content of element phytoextracted; (3) developing agricultural practices to enhance economic gains; and (4) developing techniques to recover metal(s) from plants' biomass (Chaney et al., 2021).

Hyperaccumulators, namely, *Thlaspi caerulescens*, *Salix* sp., *Sedum alfredii*, *Helianthus annuus*, and *Brassica juncea*, are reported to phytoextract Pb and Cd in various research studies (Escarré et al., 2000; Meers et al., 2007; Meighan et al., 2011; Sun et al., 2009; Zaier et al., 2010). *Alyssum murale* (Brassicaceae) and *Berkheya coddii* (Asteraceae) are regarded as universal "metal crops" (van der Ent et al., 2016).

Ultramafic soils are characterized by the presence of high concentration of trace metals (i.e. Co, Cr, Mn, and Ni). They are rich in Mg but poor in essential macronutrients (N, P, K and Ca), making them less fertile and low productive in nature (Vithanage et al., 2019). Such soils are usually

Metallophytes in Mineral Mining
DOI: https://doi.org/10.1016/B978-0-443-30110-0.00003-9

abandoned by local farmers, since they are uninhabitable for normal plants; but the fact that they host a range of endemic species makes ultramafic landscapes as potential sites for phytomining. Serpentine flora chiefly consists of metallophytes and hyperaccumulators (unusual plants capable of accumulating exceptionally high concentrations of certain elements in their biomass naturally, unaccompanied by physiological symptoms of stress), that can be a material basis for their application in phytomining and hydrometallurgy using hyperaccumulator ash (after combustion and energy production). Till date, more than 700 species of hyperaccumulator have been reported, pertaining them to be defined as plants that accumulate at least 100 mg kg^{-1} of dry weight (0.01%) of Cd or As, 1000 mg kg^{-1} (0.1%) of Co, Cu, Cr, Al, Ni, or Pb, and 10,000 mg kg^{-1} (1%) of Mn, Fe, or Zn (Sun et al., 2019a) along with bioconcentration factor and translocation factor higher than 1 (Wei et al., 2012). Different hyperaccumulators show different preferences to each metal, which can be assigned to their hyperaccumulation mechanisms, namely, hyperactivity of metal transporters and detoxification pathways (Sun et al., 2019a).

Such sites have also provided model organisms to analyze the mechanism of metal uptake by hyperactive metal transporters leading to hyperaccumulation. Studies on hyperaccumulator plants of ultramafic soils have not only led to the identification of novel physiological pathways and their intrinsic genetic mechanisms but have also led to the progression of innovative technologies like phytoremediation (Rajakaruna and Boyd, 2008), whereby the resulting crop can be subjected to phytomining.

The global market of phytoremediation reached US$6.2 billion in 2023 and is estimated to grow at a compound annual growth rate of 5.1% from 2023 to 2030 (source: Cognitive Market Research). It is anticipated to rise at a substantial rate in developed countries. Phytoextraction has emerged as the fastest-growing segment in the phytoremediation market among other phytoremediation techniques (phytostabilization, rhizodegradation, phytovolatization, phytohydraulics, and phytodegradation). The applicability of phytoextraction is dependent upon biomass production, rate of accumulation, and tolerance to toxic target metals. Exploring and engaging hyperaccumulator plants against conventional technologies offers cost-effective approach for in situ remediation, without releasing secondary pollutants. Phytomining generates revenue and gives an opportunity for exploratory research in the field of biofortification, production of biofuel, bio-ore, and biochar.

5.2 Phytomining versus agromining

Principally, both phytomining and agromining are about using plants to extract elements of commercial interest (Chaney et al., 2021). Phytomining involves degraded or mined land as part of a rehabilitation strategy, whereas agromining involves low-productivity agricultural soils and aims to generate economic returns to farmers (van der Ent et al., 2015). Agromining is a more consolidated approach in comparison to phytomining with the objective of improving ecosystem condition in ultramafic regions at landscape level. Agromining is an absolute blueprint of new agro-ecological cropping methodology and the corresponding metallurgical chain (including retrieval of phytomined elements). The first phytomining field trials were based at the US Bureau of the Mines, Reno, Nevada employing *Streptanthus polygaloides*, a naturally occurring strain of Ni hyperaccumulator and stated that a yield of 100 kg ha^{-1} of sulfur-free Ni could be produced (Chaney et al., 1998).

Phytomining can be natural (continuous) or induced (assisted), that is, driven by amendments (Liang et al., 2014). Assisted phytomining is comparable to conventional techniques in performance while being cost-effective. Upon supplementation with ammonium thiocyanate, *Brassica juncea*, *Berkheya coddii*, and *Daucus carota* have been reported to accumulate up to 20 mg kg^{-1} of gold (Prasad, 2003). Bio-assisted phytomining refers to employing specific microorganisms and plants for selective recovery of metal. Application of specific microorganisms dissolves metals from their undissolved complexes and makes them bioavailable to the hyperaccumulator plants (Maluckov, 2015). Generally, inoculation of plant growth-promoting and metal-solubilizing microbes are known to affect plants immensely by enhancing the bioavailability of target toxic metals and improving growth, biomass yield, and accumulation potential of hyperaccumulators (Kidd et al., 2018).

5.3 The rhizosphere of metallophytes

5.3.1 Physico-chemical properties of rhizosphere

Generally, in hyperaccumulation studies, the leaf receives greater attention as it provides the sequestration site for target metals in comparison to the root, which is continuously involved with metal uptake. However, the biogeochemical processes occurring at the root–soil environment are of extreme importance, as the process of metal uptake begins from the rhizosphere. Rhizosphere or the root–soil interface is the microecosystem where both biotic (root and microorganisms) and abiotic (soil and elements) components come together. The rhizosphere links plant roots and their immediate soil environment, which houses intricate physical, chemical, and biological interactions involving various microbes with specific functions, thus boosting the diversity and abundance of favorable microorganisms (He et al., 2021).

Rhizosphere is constituted by nutrient-rich zones possessing the ability to carry on extended activities for plants under harsh conditions (Chamkhi et al., 2022). Rhizosphere is a dynamic system at temporal and spatial scales, since multiple gradients coexist simultaneously. The physico-chemical conditions of the rhizospheric soil, that is, elemental concentration, bulk density, pH, moisture, exudates, microbes, redox potential, and so on, regulate the mobility and bioavailability of metals in soil (Alford et al., 2010).

In order to understand hyperaccumulation phenomenon, roots' physical characteristics along with rhizosphere chemistry and microbiology needs to be investigated. Till date, rhizospheres of only a few hyperaccumulators have been examined. A physical feature assumed to be common to most of the hyperaccumulators is a small and shallow (<0.5 m) root system with extensive fine roots contributing to metal uptake. However, this conception may disapprove upon further exploration, as herbaceous species with deep roots (~ 2 m) do exist (e.g. *Biscutella laevigata*, a Cd hyperaccumulator). The morphology of root and its growth pattern govern its access to metals present in rhizosphere. The differences in root functions among root types (primary, secondary, etc.), age, and location (zone of elongation, maturation, etc.) contribute to auxiliary variations in rhizosphere (Alford et al., 2010).

The mechanisms fundamental to mobility and immobilization may separately or jointly govern the fate of metals in soil (Zemanová et al., 2021). For example, low pH promotes reactive dissolution of metals in soil, leading to increased bioavailability (Kicińska et al., 2022); high concentration of organic matter limits the mobility of metals by virtue of its high negative charge capable of adsorbing cationic elements (Sarkar et al., 2021).

At chemical level, roots are known to alter metal solubility and bioavailability through exudates and pH changes. Chemotropism, that is, directed growth of roots toward trace elements present in soil has been recorded in some hyperaccumulators, but the mechanism is not clearly understood. Moreover, the presence of metal chaperones and chelators (phytochelatins, metallothioneins, and organic acids) may aid metal tolerance and transport within root cells. Plants regulate the availability and mobility of metals by releasing root exudates capable of enhancing solubility and forming complexes such as acetic, fumaric, malic, and oxalic acids (Lapie et al., 2019).

5.3.2 Microbiological properties of rhizosphere

Plants usually harbor tremendously variegated microbiota in different compartments, including rhizosphere, endosphere, and phyllosphere, which greatly affect its well-being and productivity (Bai et al., 2015). Various endophytic and rhizospheric bacteria constitute the plant–microecological environment.

Rhizospheric microbe–plant interactions also involve chemical processes for solubilization and chelation of metals to either promote mobility or render them immobile. Although some microbes may not appear to have any straight inclination toward root exudates and soil physicochemical properties, but still aid in detoxification. For example, *Bacillus subtilis* is shown to have high tolerance for Cd and Pb adsorption and detoxification (Li et al., 2022).

At microbiological level, the rhizospheric bacteria and fungi function as plant growth-promoting microorganisms that aid in increasing root surface area and hair formation. They may alter nutrient cycling and enhance bioavailability of elements present in rhizosphere. Rhizospheric microbes may aid in increasing soluble element transfer from soil to plant and grant increased metal tolerance.

Plant growth promoting bacteria (PGPB) enhances the plants' growth by producing 1-aminocyclopropane-1-carboxylate (ACC) deaminase, which, in turn, limits ethylene accumulation responsible for the inhibition of plant growth. Some PGPBs also produce auxins that lead to cell elongation in roots (Ma et al., 2009). Besides, PGPB also facilitate plant growth by promoting mineral acquisition such as fixing N_2, solubilizing P and other essential minerals, producing phytohormones, siderophores, and so on, and protecting the colonizing plants from a range of pathogen attack by producing antibiotics and other biocontrol agents. PGPB also play a role in stress alleviation by synthesizing ACC deaminase and increasing tolerance to drought, salinity, and metal-induced stress along with reducing ethylene-mediated stress. PGPB may gain entry in the plant xylem upon infection of plant roots and become endophytes within; however, this route is not validated in hyperaccumulators (Rajkumar et al., 2009). Endophytic bacteria present in roots and shoots of hyperaccumulatorplants can be similar soil bacteria as in *Alyssum bertolonii* (Barzanti et al., 2007), or different as in *T. caerulescens* and Thlaspi *goesingense* (Idris et al., 2004; Lodewyckx et al., 2002). Rhizosphere isolates extracted from *T. goesingense* and *Alyssum serpyllifolium* (Idris et al., 2004; Ma et al., 2009) contained siderophores, which are characterized by low molecular weight, water solubility, high-affinity Fe-chelating compounds that may enhance the availability of Fe and other trace elements possibly. The rhizospheres of hyperaccumulators may be a niche for metal-resistant PGPB compared to bulk soil. The content of phosphorous has been reported to be higher in rhizospheric soil in comparison with non-rhizospheric soil, validating the presence of some chemical substance in rhizosphere, which solubilizes unavailable phosphates (Ca, Fe, and Al) and make P available (Adejumo et al., 2018).

Arbuscular mycorrhizal associations have been reported in several hyperaccumulator plant species except from the ones that belong to Brassicaceae, as it is generally a nonmycorrhizal family. Although the factors contributing to the rate of mycorrhizal colonization in hyperaccumulators are unknown to a large extent, the glucosinolates are responsible for its inhibition in Brassicaceae (Pongrac et al., 2008). Fungal species other than mycorrhizae are also reported in hyperaccumulator rhizospheres. The role these microorganisms play in hyperaccumulation phenomenon is unclear yet, but their ability to sequester and volatilize trace elements are clearly noted (Wangeline, 2007). Research shows that certain dicot herbaceous plants belonging to Brassicaceae and Caryophyllaceae, which do not show associations with mycorrhizae, require a longer evolution time to develop heavy metal resistance than in grasses (Ernst, 2001).

5.4 Role of rhizobiome in hyperaccumulation

Soil is a complex ecosystem involving diverse microbes vital to maintain soil fertility and plant productivity via interplay among physical, chemical, and biological components (Fig. 5.1). Additionally, root network creates a rhizospheric microbial community composed of bacteria, fungi, algae, and protozoa capable of modulating the bioavailability and mobility of toxic elements (Asare et al., 2023).

The rhizobiome, a subset of specialized microorganisms that colonize the plant rhizosphere and endosphere give pivotal contribution to the

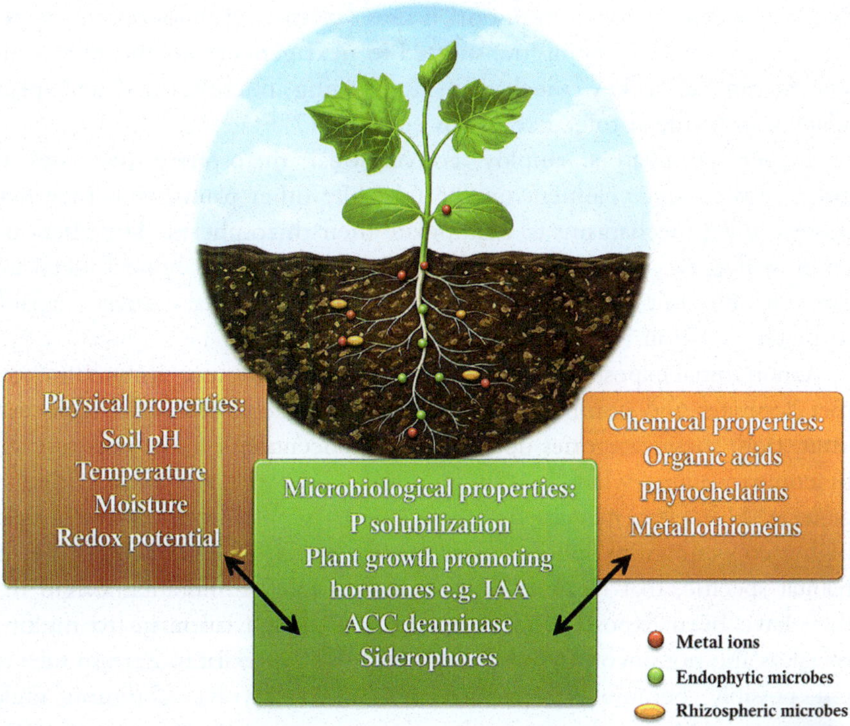

Figure 5.1 Interplay between physical, chemical, and microbiological properties of soil.

hyperaccumulator phenotype. The interpretation of hyperaccumulator rhizobiomeis is much required to implement its biotechnological applications in plant-based remediation strategies. Research has shown that the transfer of endophytes and rhizobacteria of hyperaccumulators to non-accumulators may improve plant biomass, root proliferation pattern, metal uptake potential, translocation, and accumulation (Visioli et al., 2015a). Both kinds of bacteria (rhizospheric and endophytic) show similar growth-promoting behavior and mechanisms to metal tolerance; but endophytic bacteria experience lesser competitive environment, since they reside in host tissues and are affected by plant health conditions, stage of growth, and nutritional status; hence, favorable effects are readily elicited (Kong and Glick, 2017). However, rhizobacteria are exposed to changing soil conditions that regulate their proliferation and functioning, exposing them to variations that endophytes are unaware of. Some endophytic bacterial species colonize transport tissues (xylem and phloem) and hence are at a possibility of being transmitted to next generations through seeds, which can be a key mechanism underlying microbe-mediated plant adaptation to metal tolerance (Thijs et al., 2017).

Hyperaccumulators employ conventional rhizosphere processes to enhance their trace element uptake (just like other plants), but they may possess novel mechanisms to manipulate their rhizospheres. The phenomenon of hyperaccumulation is not just dependent on mobilized metal ion bioavailability but also on the presence of plant/microbial exudates capable of metal chelation.

Upon initial exposure to metals, a similar reaction is encountered in both metallophytes and nonmetallophytes with an exception in their tolerance limit. Both these categories uptake higher concentration of metals from soil in comparison to when grown in normal soil and face similar kind of damage. However, metallophytes are able to withstand this stress both in higher concentrations and in longer tenures better. The tolerance is usually habitat specific, that is, metallophytes are capable of tolerating metals that they have been exposed and evolved to. Although tolerance to multiple metals is also not uncommon, for example, plants tolerant to Zn also tolerate Cu possibly because similar divalent cations involve common metal transporters. Metallophytes involve different strategies to survive under heavy metal stress. Some plants become excluders to reduce the uptake of heavy metals. Their roots exude organic acids, which, in turn, chelates heavy metals, making them incapable of moving across cell membranes. If heavy metals are able to surpass root membrane and get translocated to shoot, they get immobilized and compartmentalized and thus their concentration in cytosol gets reduced. Excluders are in contrast to hyperaccumulators, which are able to accumulate metals multiple folds higher than present in soil. Rhizospheric microorganisms play pivotal roles in survival of plants by manipulating the bioavailability of metals in rhizosphere.

Microbesin metalliferous soils generally develop metal resistance, but this feature is not seen in microbes inhabiting normal soils. Moreover, endophytic bacteria residing in inner tissues of hyperaccumulators, such as root cortex, may have been better adapted to tolerate toxic metal concentrations in comparison to rhizobacteria. The distribution of endophytic microbes in different tissues and organs somehow explains variability in metal tolerance by different parts of the plant (He et al., 2013).

The ability of rhizospheric microbiomes to withstand high content of heavy metals is principle to induce accumulation and tolerance during phytoremediation. For example, siderophores, carboxylic acids, and phosphate solubilizers secreted by PGPBs can aid in dissolution of As, making it bioavailable for plant uptake (Cabello-Conejo et al., 2014). In *Pteris vittata* plants growing under As stress, the inoculation of PGPBs (namely, *Bacillus* sp., *Delftia* sp., *Pseudomonas* sp., *Pseudoxanthomonas* sp., and *Variovorax* sp.) led to improved plant biomass and better efficiency in the removal of As in contaminated soils (Yang et al., 2012). Cd bioavailability was reported to be increased, and its absorption was facilitated in plant roots via siderophores, phosphate solubilizers, antibiotics, and hydrocyanic and indoleacetic acids isolated from PGPB (He et al., 2020). Microbial populations and host plants frequently exhibit positive relationship with each other. Metal-polluted soil results in development of resistance in inhabiting rhizobacteria (González Henao and Ghneim-Herrera, 2021). Hyperaccumulators can significantly modify the ecosystem of rhizosphere and affect the microbial community to remediate pollutants. For example, in *S. alfredii*, a Cd-stressed hyperaccumulator, the actinobacterial genus *Streptomyces* led to most of the divergence of bacterial community composition in the rhizosphere of hyperaccumulating ecotype in comparison to non-hyperaccumulating ecotype (Hou et al., 2018). In *Triticum aestivum*, the bacterial phyla Acidobacteria, Bacteroidetes, Deltaproteobacteria, and Gemmatimonadetes were specifically found to be enriched in the rhizosphere of high-Cd-accumulating wheat genotypes, where they participate in metal activation, thus promoting Cd uptake (Lu et al., 2023).

In metalliferous soils, native microorganisms, which also include several species of PGPB, have evolved efficiently to avoid the toxicity generated by heavy metals. These mechanisms include (1) biosorption (adsorption of metal ions on the cell wall), (2) bioprecipitation (conversion of free metal into insoluble complex), (3) exclusion (active efflux of metal ions outside the microbial cell surface), (4) bioaccumulation (uptake of metal ions inside the cell via active or passive transport), and (5) biotransformation (conversion of toxic metals to lesser toxic forms). This is the reason why strains collected from severely metal polluted sites show better tolerance than those from unpolluted sites.

Application of PGPB with metal detoxifying traits is a cost-effective, sustainable, and ecologically sound approach for phytomining. Metal-resistant PGPB when applied as bioinoculant or biofertilizer not only improves the growth of plants in metal stressed soil but also better strategize in evading toxicity generated.

Actinobacteria, specifically *Streptomyces* and *Pseudomonas*, are examples of PGPBs that have been reported to improve plant growth in highly contaminated soil (Pb and Zn) (Hołtra and Zamorska-Wojdyła, 2020). The rhizosphere and endosphere of metallophyte*Commelina communis* grown on a site with high concentration of Cu was found to be to be significantly enriched in *Sphingomonas* and *Pseudomonas*. Both these genera were resistant to heavy metal and shown to be ecologically beneficial along with possessing multiple plant growth-promoting traits and possess solubilizing ability for heavy metals (He et al., 2021). *Sphingomonas* sp. has also been shown to colonize rhizospheric soil of *Zea mays* and protecting them against toxicity in soil heavily contaminated with Cu (Sheng et al., 2012). A strain of *Sphingomonas* isolated from hyperaccumulator *S. alfredii*, was shown to enhance plant growth and provide protection to its roots against Cd damage (Chen et al., 2014). Another strain of *Sphingomonas* promoted Cd translocation and accumulation in *Brassica napus* shoots (Pan et al., 2017). *Pseudomonas*, a well-known plant growth-promoting rhizobacterium and its several species were isolated from metal-rich environments. For instance, Cu-tolerant *Pseudomonas putida* improved the growth of *Elsholtzia splendens*, a Cu-tolerant plant, significantly and enhanced Cu translocation as well as accumulation (Xu et al., 2015). *Pseudomonas koreensis*, isolated from roots of *Miscanthus sinensis* in mine tailings, is highly tolerant to As, Cd, Pb, and Zn. Furthermore, inoculation of *P. koreensis* improved *M. sinensis* growth by alleviating toxicity generated by heavy metals through production of indole-3-acetic acid (IAA) and 1-aminocyclopropane-1-carboxylate deaminase (ACC deaminase), which led to enhanced activity of antioxidant enzymes along with reduction in lipid peroxidation (Babu et al., 2015).

Pseudomonas was reported to be the most dominant genus in shoots and roots of two hyperaccumulators, *Arabis alpine* and *Dysphania ambrosioides*, upon investigating their endophytic bacterial community. *Microbacterium* sp., an endophyte isolated from from *A. alpine* was reported to enhance host plants' growth significantly in the presence of multiple heavy metals (Sun et al., 2019b). Rhizospheric microbes, namely, *Acinetobacter, Bacillus, Gluconacetobacter*, and *Pseudomonas*, have been reported to influence Zn bioavailability by modulating pH and root growth (Costerousse et al., 2018). In *Salix* sp. mycorrhizal fungal association contributes to tolerance against excess Cd and Zn (Hrynkiewicz et al., 2012). In the rhizosphere of *Brassica napus* L., *Enterobacter* sp., and *Klebsiella* sp. enhance the accessibility of nutrients as well as high metals under Cd, Pb, and Zn stress (Jing et al., 2014).

Hyperaccumulator plants can sequester heavy metals via exudates that modulate soil pH and promote mobilization of metals and detoxification

(Das et al., 2017). Hyperaccumulator ecotypes of plants, namely, *S. alfredii*, show varied microbial community compositions in comparison to non-accumulator types. Rhizospheres of these ecotypes exhibit enhanced enzymatic activity and reduced metal content in soil (Hou et al., 2017). A summary of relevant studies on microbes and their effects on plants applied for phytoremediation or phytomining has been presented in Table 5.1.

5.5 Biochemical and molecular mechanisms of microbe-assisted phytomining

The activities at rhizosphere regulate the response of different plants (hyperaccumulator, excluder, tolerant, and indicator) under heavy metal stress. The exudates secreted by hyperaccumulator roots enhance the solubility and dissociation of metal ions in soil solution in contrast to excluders, which render these ions insoluble. Rhizospheric activities are efficient enough to induce alterations in chemistry of soil solution by modulating pH and ionic strength, which therefore affects the solubilization of metals.

Soil pH determines the sorption processes of metals in the soil. The bioavailability of metals is ascertained by the concentration of metal ions available in water-soluble and exchangeable fractions. Ligands, such as, phytosiderophores, citrate, oxalate, etc. present in soil, affect the concentration of metals in soil solution by binding and forming metal complexes thereby enhancing their sequestration in plant cells (Seshadri and Bolan, 2015).

An approach to metal avoidance in plants is loading of metal ions in cell wall (Krzeszowska et al., 2022). The carboxylic groups of polygalacturonic acids of pectin present in cell wall can bond with cationic metal ions and avert free metal ions from entering plant cells.

Mycorrhizae in rhizosphere alters the pH of soil by releasing organic acids and affects the availability of heavy metals. Heavy metal mobility in soil is regulated by the components of organometallic complexes thus formed. Humus-poor but metal-rich soil is difficult to be colonize,d as organically bound metals (in conjunction with humic acids) are less taken in comparison to free ions (Leuschner and Ellenberg, 2017).

During root expansion, plants are known to exude carbon compounds with low molecular weight, such as sugars and organic acids, actively or

Table 5.1 Summary of relevant studies on microbes and their effects on plants applied for phytoremediation or phytomining.

Microbe	Plant	Effect	References
Serratia sp. AI001 and *Klebsiella* sp. AI002	*Solanum nigrum*	Increase in plant biomass and Cd translocation and can be attributed to IAA production	Ullah et al. (2022)
Jeotgalicoccus huakuii, and *Bacillus amyloliquefaciens*	*Cynodon dactylon*, *Eleusine indica*	Increase in Hg accumulation and can be attributed to IAA production	Ustiatik et al. (2022)
Pseudomonas sp.	*Medicago sativa*	Increase in plant biomass and Cr root concentrations, can be attributed to N fixation, P solubilization, production of IAA, and siderophores	Tirry et al. (2021)
PGPE consortium	Intercropped: *Sedum alfredii* and *Vicia fava*	Increase in plant biomass and Cd and Pb uptake	Tang et al. (2020)
Serratia marcescens	*Pteris vittata*	Increase uptake of Cd, Cr, and V, can be attributed to ACC deaminase activity, production of siderophore, IAA, and P solubilizing activity	Wang et al. (2018)
Buttiauxella sp. SaSR1 PGPE	*S. alfredii*	Increase in plant growth and Cd accumulation	Wu et al. (2018)
Kocuria sp. CRB15	*Brassica nigra*	Increase in plant growth and can be attributed to IAA production and P solubilization	Hansda et al. (2017)
Streptomyces pactum	*Sorghum bicolor*	Increase in biomass and uptake of Zn, Pb, and Cu and can be attributed to the production of IAA, siderophores, and ACC deaminase activity	Ali et al. (2017)

Pantoea stewartii ASI11, *Enterobacter* sp. HU38, and *Microbacterium arborescens* HU33	*Leptochloa fusca*	Increase in plant biomass and U and Pb uptake	Ahsan et al. (2017)
Pseudomonas azotoformans ASS1	*Trifolium arvense*	Increase in Cu, Ni, and Zn contents and, can be attributed to production of IAA, siderophore, ACC deaminase, and P solubilization	Ma et al. (2017)
Chryseobacterium humi ECP37, *Pseudomonas reactans* EDP28, and *Pseudomonas fluorescens* S3X	*Zea mays*	Increase in plant biomass and Cd uptake and can be attributed to production of ACC deaminase, siderophores, and P solubilization	Moreira et al. (2016)
Variovorax NB24	*Noccaea tymphaea* and *Bornmuellera tymphaea*	Increased Ni accumulation in roots and shoots and can be attributed to the production of ACC deaminase and IAA compounds	Durand et al. (2016)
Bacillus pumilus E2S2 and *Bacillus* sp. E1S2	*Sedum plumbizincicola*	Increase in plant biomass and Cd and Zn accumulation and can be attributed to the production of IAA, siderophores, ACC deaminase, and P solubilization	Ma et al. (2015)
Arthrobacter sp. Ncr-1 and *Microbacterium* sp. Ncr- 8	*Noccaea caerulescens*	Increase in biomass and Ni uptake along with Fe, Co, and Cu and can be attributed to the production of IAA and ACC deaminase	Visioli et al. (2015b)
Sphingomonas SaMR12	*S. alfredii*	Increase in plant biomass and Zn accumulation	Chen et al. (2014b)

(*Continued*)

Table 5.1 (Continued)

Microbe	Plant	Effect	References
Enterobacter sp. CBSB1	*Brassica juncea*	Increase in plant biomass and Cd and Pb accumulation	Qiu et al. (2014)
Pseudomonas sp. Lk9	*Solanum nigrum*	Increase in Cd, Cu, and Zn attributed to the bacterial production of organic acids, siderophores, and biosurfactants	Chen et al. (2014a)
Rahnella sp. JN27	*Amaranthus hypochondriacus* and *Amaranthus mangostanus*	Increase in plant biomass and Cd uptake	Yuan et al. (2014)
Microbacterium sp., *Arthrobacter* sp., and *Streptomyces* sp.	*Alyssum pintodasilvae*	Increase in plant biomass and Ni accumulation and attributed to the bacterial production of organic acids, siderophores, IAA compounds, and P solubilization	Cabello-Conejo et al. (2014)
Bacillus megaterium var. *phosphaticum*	*Trifolium hybridum*, *Alopecurus pratensis*, *Poa pratensis*, *Hordeum violaceum*, *Ranunculus kotschyi*, and *Cerastium* sp.	Increase in uptake of Pb, Ni, B, Mn, and Zn	Gullap et al. (2014)

passively in the proximity of root hair (Canarini et al., 2019), and modulate microbial community (Hu et al., 2018) via enzyme activities (Zhang et al., 2020), which are known to affect solubility and plants' accessibility to heavy metal ions.

Root exudations of hyperaccumulators aid in the selection of microbes. Microbial stimulation via chemotaxis and secretion of flavonoids is known to provide a harbor for colonization of root surface and contribute to uptake of metal ions upon solubilization. The pace of production of exudates by roots of hyperaccumulator and the affinity of rhizospheric microorganisms influences metal availability and accumulation significantly, further improving hyperaccumulators' phytoextraction potential.

Organic acids are involved in trapping and chelating the metal ion, thereby leading to detoxification. For example, in leaf tissues of *T. goesingense*, citrate has been shown to bind Ni; malate has been found to bind Zn and Cd (in *Arabidopsis halleri* and *T. caerulescens*, respectively). Amino acids play a very important role in forming stable complexes with bivalent cations in hyperaccumulators, aiding mainly in sequestration of metal ions both in cell and xylem sap. For instance, high content of histidine is seen in the roots of Ni hyperaccumulators, suggesting its involvement in Ni hyperaccumulation. Proline is found to be accumulated under heavy metal stress, as it inhibits free radical formation along with maintaining reducing environment by increasing glutathione levels, aiding in cellular defense against metal toxicity (Singh et al., 2016). An increased concentration of glutathione synthetase is found in plants under heavy metal stress. Alpha-tocopherol is involved in scavenging reactive oxygen species and lipid peroxides, which happens to be the most active form of vitamin E. Its biosynthesis takes place in higher plants' plastids, and its increased concentration is found in plants growing under heavy metal stress.

Chelation implies binding of cysteine-rich peptides with metal ions, leading to the formation of nontoxic complexes that are stored in vacuoles to check free movement of the latter in the cytosol. Phytochelatins and metallothioneins are the most salient cysteine-rich peptides that act as chelators. Phytochelatins are glutamic acid, cysteine-, and glycine-containing peptides possessing low molecular weight. Their biosynthesis from glutathione via phytochelatin synthase takes place in response to metal stress. Since they show high affinity to heavy metals, they are potential biomarkers for the identification of early stress conditions. Phytochelatins are the best characterized chelators in plant system and are vital for metal ion homeostasis, besides detoxifying heavy metals, and hence regulate the bioavailability in plant cells. The biosynthesis and activation of metallothioneins also happens during heavy metal stress conditions. Study on Noccaea *caerulescens* revealed higher levels of metallothioneins in comparison to *Arabidopsis thaliana*, making it more Cd-tolerant and also showing better accumulation (Jain et al., 2018). Exudation of glutathione derivatives and phytochelatins in response to As stress in *Lupinus albus* contributes to active As efflux in plants (Frémont et al., 2024).

Heavy metals are known to interact with nucleic acids and protein in living organisms, rendering them inactive. In turn, plants are known to activate chaperone proteins, which bind to excess metals and are also capable of protecting and repairing cellular machinery during stress (Jain et al., 2018). Nicotinamine is a nonprotein amino acid that binds to heavy metals when present in excess concentrations and aids in metal homeostasis. Roots of *A. halleri* were shown to exude high levels of nicotianamine to form complex with Zn and impart tolerance (Tsednee et al., 2014). In the dissolved organic matter within the rhizosphere of *S. alfredii*, elevated contents of organic acids and mobilized Zn were seen in comparison to its nonaccumulator accessions (Li et al., 2012). Several nicotianamine synthase genes are upregulated in *A. halleri*, *N. caerulescens*, and hyperaccumulator accessions in *S. alfredii* (Deinlein et al., 2012; Liang et al., 2014).

Since most hyperaccumulated metals happen to be essential micronutrients (Cu, Mn, Ni, and Zn) for the plant, specific transporters are present in the root membrane. In addition to transporters at root membranes, mechanisms such as reduced sequestration in root vacuoles are present to promote translocation to shoot. The uptake of metal ions takes place through diffusion into the root epidermis to some extent, but the passage occurs along the apoplastic and symplastic pathway primarily (Yan et al., 2020). The activity of transporters present in root cell membranes is responsible for metal uptake and provides entry to symplastic pathway, which is effectively controlled by metal carriers or complexing agents. Or else, the elements may find entry into the apoplast and later find way to symplast via transporters present in endodermis around vascular cylinders. Several transporter proteins are also employed in uptake and translocation of metals in apoplastic pathway, whereas the movement of toxic elements like Pb is restricted by casparian strips. Lignification of endodermis ensures that no metal effluxes out of root vasculature (Asare et al., 2023). The uptake, transport, detoxification, and sequestration of elements is very tightly regulated, and the underlying mechanisms are key in differentiating nonaccumulators from hyperaccumulators.

5.6 Conclusion

Phytoremediation implies extracting/stabilizating toxic contaminants for safe disposal, whereas phytomining implies to reclaiming precious metals from cultivation of hyperaccumulator plants for monetary return (McGrath and Zhao, 2003). Phytomining provides the opportunity of extracting ores/mineralized soils, which are otherwise uneconomical for exploration conventionally. Further, bio-ores offer the advantage of being richer than conventional ores in terms of metal concentration, also occupying lesser space for storage because of lower density. Phytomining agro-ecosystem is of value for sustainable rural development for the welfare of local community. Thus, phytomining abets revenue generation by reclaiming valuable metals (Au, Cd, Co, Cu, Mn, Ni, Pb, Se, Tl, U, and Zn), facilitates rehabilitation, and leads to the closure of mining sites sustainably (Wilson-Corral et al., 2011). The global phytomining market is approximately around US$34–54 billion (Evangelou and Deram, 2014).

Although agromining/phytomining technology has been aptly illustrated in academia, an extensive on-field implementation is lacking, which shows lack of technological awareness in the minerals' industry. The effectiveness of hyperaccumulator plants and efficacy of phytomining relies on several biogeochemical factors pertaining to (1) plant, that is, biomass production, accumulation, metal tolerance, and resilience to both flood and drought conditions; (2) soil, that is, moisture, pH, salinity, nutrient availability, metal speciation and bioavailability, rhizobiological activity, release, and competing ions influencing plant growth and metal bioavailability in the soil–water system; and (3) environment, that is, prevailing temperature, humidity, and water stress (Ali et al., 2013; Mahar et al., 2016).

The economic feasibility of phytomining operation is determined by factors such as the soil metal concentration, plant metal accumulation, annual production of biomass, and whether or not the biomass generated can be used as biofuel. However, the most crucial factor of all is the value of metal being phytomined globally (Harris et al., 2009). In this context, Au, Tl, Co, and Ni are the best suitable candidate metals for phytomining for their high market prices and accumulation concentration. Although, U is much valuable than most other metals, but its reported accumulation concentration (100 mg kg^{-1}) in biomass (10000 kg ha^{-1}) of *Atriplex confertifolia* is low, rendering its phytomining commercially unprofitable (Mahar et al., 2016; Sheoran et al., 2009).

Phytomining is environmentally responsible, less intrusive, esthetically pleasing, safe, and nondestructive technology for on-site remediation, with high rates of acceptance publically and commercially. A well-designed phytoremediation/phytomining project will yield an economically feasible metal enriched bio-ore. Thereafter, the mined soil is possibly put to agricultural or commercial use (Sheoran et al., 2009). Practicing phytomining will minimize the risk of transfer of metals via surface runoff or wind and the possibility of leaching into aquifers and also check soil erosion at the same time. Plants employed for phytomining (or agromining) are hugely benefitted by synchronization with microorganisms. The potential of microbes in amplifying plant metal yields is recognized in several phytomining-related studies (Alves et al., 2020; Benizri and Kidd, 2018) and has been discussed in this chapter.

5.7 Future recommendations

The mechanisms underlying metal uptake, translocation, sequestration, tolerance, and hyperaccumulation in candidate plant species and their parts require further investigation and need to be discoursed at the molecular level. The identification of genes responsible for hyperaccumulation and its introduction/upregulation in candidate plants for improved phytomining needs to be carried out vigorously. The application of biotechnology and genetic engineering for optimizing phytomining operations cannot be disregarded. Transfer of genes responsible for enhanced metal uptake, translocation, sequestration, and tolerance into desired candidates is a potential strategy for designing genetically modified hyperaccumulator plants. At the same time, selection of individual plant species with genes coding for enhanced metal uptake, improved biomass production and superior metal tolerance is in itself a broad area for research.

The physical, chemical, and biological components of rhizosphere need advanced studies in order to optimize phytomining. Extensive and meticulous investigation is required in the field of microbe-assisted phytomining to aid the rehabilitation of minelands along with revenue generation. Exploring the role of phytohormones (IAA, cytokinins, gibberellic acid, indolebutyric acid, and naphthylacetic acid) along with rhizobacteria provides a new research dimension regarding mechanisms of metal uptake and their stabilization for sustainable remediation of polluted sites.

The setting up of hyperaccumulators' seed bank needs optimization in order to implement phytomining projects in various ecological zones. The research outcomes of various ecological zones can further assist in designing new projects, which will be better suitable for the remediation of pollutants. Before the introduction of exotic invasive species for phytomining, measures for protecting and conserving native plant species and biodiversity must be implemented.

The harvested biomass must be subjected to drying and incineration for the generation of a high-grade bio-ore. Special considerations to avoid transfer of heavy metals via food chain (soil to plant to animal to human) must be kept. Under no circumstances shall the harvested biomass put to be use other than as bio-ore. From economic point of view, the conversion of biomass generated by hyperaccumulator plants into biofuel, bio-ore, and biochar requires investment and technical expertise. Cost–benefit analysis over time period consumed needs to be determined for a final conclusion.

Field trials are immensely important to come up with "real" results without modified variables. Aspects, namely, compatibility of the host plants with microorganisms; their performance and proliferation in contaminated soil needs to be investigated under natural field conditions over longer periods of time. It is visualized that microbial inoculations may begin to replace usage of other amendments (inorganic fertilizers, phytohormones, and chelators) in phytoremediation/phytomining and subsequently reduce the associated costs.

References

Adejumo, S.A., Tiwari, S., Shinde, V., Sarangi, B.K., 2018. Heavy metal (Pb) accumulation in metallophytes as influenced by the variations in rhizospheric and non-rhizospheric soils physico-chemical characteristics. International Journal of Phytoremediation 20 (3), 237–248. Available at https://doi.org/10.1080/15226514.2017.1374333, http://www.tandf.co.uk/journals/titles/15226514.asp.

Ahsan, M.T., Najam-ul-haq, M., Idrees, M., Ullah, I., Afzal, M., 2017. Bacterial endophytes enhance phytostabilization in soils contaminated with uranium and lead. International Journal of Phytoremediation 19 (10), 937–946. Available at https://doi.org/10.1080/15226514.2017.1303813, http://www.tandf.co.uk/journals/titles/15226514.asp.

Alford, E.R., Pilon-Smits, E.A.H., Paschke, M.W., 2010. Metallophytes-a view from the rhizosphere. Plant and Soil 337 (1), 33–50. Available at https://doi.org/10.1007/s11104-010-0482-3.

Ali, H., Khan, E., Sajad, M.A., 2013. Phytoremediation of heavy metals: concepts and applications. Chemosphere 91 (7), 869–881. Available at https://doi.org/10.1016/j.chemosphere.2013.01.075, https://www.elsevier.com/locate/chemosphere.

Ali, A., Guo, D., Mahar, A., Ma, F., Li, R., Shen, F., et al., 2017. Streptomyces pactum assisted phytoremediation in Zn/Pb smelter contaminated soil of Feng County and its impact on enzymatic activities. Scientific Reports 7 (1), 46087. Available at https://doi.org/10.1038/srep46087.

Alves, A.R.A., Oliveira, R.S., Silva, E.F., Novo, L.A.B., 2020. Role of bacteria and mycorrhizal fungi in phytomining: status and future perspectives. Microbe Mediated Remediation of Environmental Contaminants. Elsevier, Portugal, pp. 15–26. Available at https://doi.org/10.1016/B978-0-12-821199-1.00002-X, https://www.elsevier.com/books/microbe-mediated-remediation-of-environmental-contaminants/kumar/978-0-12-821199-1.

Asare, M.O., Száková, J., Tlustoš, P., 2023. Mechanisms of As, Cd, Pb, and Zn hyperaccumulation by plants and their effects on soil microbiome in the rhizosphere. Frontiers in Environmental Science 11, 1157415. Available at https://doi.org/10.3389/fenvs.2023.1157415, https://www.journal.frontiersin.org/journal/environmental-science.

Babu, A.G., Shea, P.J., Sudhakar, D., Jung, I.B., Oh, B.T., 2015. Potential use of Pseudomonas koreensis AGB-1 in association with Miscanthus sinensis to remediate heavy metal(loid)-contaminated mining site soil. Journal of Environmental Management 151, 160–166. Available at https://doi.org/10.1016/j.jenvman.2014.12.045, https://www.sciencedirect.com/science/journal/03014797.

Bai, Y., Müller, D.B., Srinivas, G., Garrido-Oter, R., Potthoff, E., Rott, M., et al., 2015. Functional overlap of the Arabidopsis leaf and root microbiota. Nature 528 (7582), 364–369. Available at https://doi.org/10.1038/nature16192, http://www.nature.com/nature/index.html.

Barzanti, R., Ozino, F., Bazzicalupo, M., Gabbrielli, R., Galardi, F., Gonnelli, C., et al., 2007. Isolation and characterization of endophytic bacteria from the nickel hyperaccumulator plant *Alyssum bertolonii*. Microbial Ecology 53, 306–316.

Benizri, E., Kidd, P.S., 2018. The role of the rhizosphere and microbes associated with hyperaccumulator plants in metal accumulation. In: van der Ent, A., Echevarria, G., Baker, A.J.M., Morel, J.L. (Eds.), Agromining: Farming for Metals. Mineral Resource Reviews. Springer, Cham, pp. 157–188. Available at https://doi.org/10.1007/978-3-319-61899-9_9.

Cabello-Conejo, M.I., Becerra-Castro, C., Prieto-Fernández, A., Monterroso, C., Saavedra-Ferro, A., Mench, M., et al., 2014. Rhizobacterial inoculants can improve nickel phytoextraction by the hyperaccumulator Alyssum pintodasilvae. Plant and Soil 379 (1–2), 35–50. Available at https://doi.org/10.1007/s11104-014-2043-7.

Canarini, A., Kaiser, C., Merchant, A., Richter, A., Wanek, W., 2019. Root exudation of primary metabolites: Mechanisms and their roles in plant responses to environmental stimuli. Frontiers in Plant Science 10, 422679. Available at https://doi.org/10.3389/fpls.2019.00157.

Chamkhi, I., El Omari, N., Balahbib, A., El Menyiy, N., Benali, T., Ghoulam, C., 2022. Is: the rhizosphere a source of applicable multi-beneficial microorganisms for plant enhancement? Saudi Journal of Biological Sciences 29 (2), 1246–1259. Available at https://doi.org/10.1016/j.sjbs.2021.09.032, http://www.ksu.edu.sa/sites/Colleges/CollegeofScinces/BotanyDepartment/SJBS/Default.aspx.

Chaney, R.L., Angle, J.S., Baker, A.J.M., Li, J.M., 1998. Method for phytomining of nickel, cobalt, and other metal from soil. US Patent 5-1998-711–784

Chaney, R.L., Baker, A.J.M., Morel, J.L., 2021. The long road to developing agromining/phytomining. In: van der Ent, A., Baker, A.J., Echevarria, G., Simonnot, M.O., Morel, J.L. (Eds.), Agromining: Farming for Metals. Mineral Resource Reviews. Springer, Cham. Available at https://doi.org/10.1007/978-3-030-58904-2_1.

Chen, L., Luo, S., Li, X., Wan, Y., Chen, J., Liu, C., 2014a. Interaction of Cd-hyperaccumulator Solanum nigrum L. and functional endophyte Pseudomonas sp. Lk9 on soil heavy metals uptake. Soil Biology and Biochemistry 68, 300–308. Available at https://doi.org/10.1016/j.soilbio.2013.10.021.

Chen, B., Shen, J., Zhang, X., Pan, F., Yang, X., Feng, Y., et al., 2014b. The endophytic bacterium, Sphingomonas SaMR12, improves the potential for zinc phytoremediation by its host, Sedum alfredii. PLoS One 9 (9), e106826. Available at https://doi.org/10.1371/journal.pone.0106826.

Chen, B., Zhang, Y., Rafiq, M.T., Khan, K.Y., Pan, F., Yang, X., et al., 2014. Improvement of cadmium uptake and accumulation in Sedum alfredii by endophytic bacteria Sphingomonas SaMR12: effects on plant growth and root exudates. Chemosphere 117 (1), 367–373. Available at https://doi.org/10.1016/j.chemosphere.2014.07.078, https://www.elsevier.com/locate/chemosphere.

Costerousse, B., Schönholzer-Mauclaire, L., Frossard, E., Thonar, C., 2018. Identification of heterotrophic zinc mobilization processes among bacterial strains isolated from wheat rhizosphere (Triticum aestivum L). Applied and Environmental Microbiology 84 (1). Available at https://doi.org/10.1128/AEM.01715-17, http://www.aem.asm.org/content/84/1/e01715-17.full.pdf. e01715-177.

Das, S., Chou, M.L., Jean, J.S., Yang, H.J., Kim, P.J., 2017. Arsenic-enrichment enhanced root exudates and altered rhizosphere microbial communities and activities in hyperaccumulator Pteris vittata. Journal of Hazardous Materials 325, 279–287. Available at https://doi.org/10.1016/j.jhazmat.2016.12.006, www.elsevier.com/locate/jhazmat.

Deinlein, U., Weber, M., Schmidt, H., Rensch, S., Trampczynska, A., Hansen, T.H., et al., 2012. Elevated nicotianamine levels in Arabidopsis halleri roots play a key role in zinc hyperaccumulation. The Plant Cell 24 (2), 708–723. Available at https://doi.org/10.1105/tpc.111.095000, http://www.plantcell.org/content/24/2/708.full.pdf+html.

Durand, A., Piutti, S., Rue, M., Morel, J.L., Echevarria, G., Benizri, E., 2016. Improving nickel phytoextraction by co-cropping hyperaccumulator plants inoculated by plant growth promoting rhizobacteria. Plant and Soil 399 (1-2), 179–192. Available at https://doi.org/10.1007/s11104-015-2691-2.

Echevarria, G., Baker, A., Benizri, E., 2015. Agromining for nickel: a complete chain that optimizes ecosystem services rendered by ultramafic landscapes. In: Proceedings of the 13th SGA Biennial Meeting on Mineral Resources in a Sustainable World, Nancy, France, vol. 1–5, pp. 1465–1467

Ernst, W.H.O., 2001. Evolutive Anpassungsmechanismen an Schwermetallböden. In: Larcher, W. (ed.), Ökophysiologie der Pflanzen. Stuttgart, E. Ulmer. pp. 353–355.

Escarré, J., Lefèbvre, C., Gruber, W., Leblanc, M., Lepart, J., Rivière, Y., et al., 2000. Zinc and cadmium hyperaccumulation by Thlaspi caerulescens from metalliferous and nonmetalliferous sites in the Mediterranean area: implications for phytoremediation. New Phytologist 145 (3), 429–437. Available at https://doi.org/10.1046/j.1469-8137.2000.00599.x.

Evangelou, M.W.H., Deram, A., 2014. Phytomanagement: a realistic approach to soil remediating phytotechnologies with new challenges for plant science. International Journal of Plant Biology & Research 2 (4), 1023.

Frémont, A., Sas, E., Sarrazin, M., Brisson, J., Pitre, F.E., Brereton, N.J.B., 2024. Arsenic stress triggers active exudation of arsenic–phytochelatin complexes from Lupinus albus roots. Journal of Experimental Botany 75 (18), 5897–5908. Available at https://doi.org/10.1093/jxb/erae272, http://jxb.oxfordjournals.org/.

González Henao, S., Ghneim-Herrera, T., 2021. Heavy metals in soils and the remediation potential of bacteria associated with the plant microbiome. Frontiers in Environmental Science 9, 604216. Available at https://doi.org/10.3389/fenvs.2021.604216.

Gullap, M.K., Dasci, M., Erkovan, H.İ., Koc, A., Turan, M., 2014. Plant growth-promoting rhizobacteria (PGPR) and phosphorus fertilizer-assisted phytoextraction of toxic heavy metals from contaminated soils. Communications in Soil Science and Plant Analysis 45 (19), 2593–2606. Available at https://doi.org/10.1080/00103624.2014.929702, http://www.tandfonline.com/loi/lcss20.

Hansda, A., Kumar, V., Anshumali, 2017. Cu-resistant Kocuria sp. CRB15: a potential PGPR isolated from the dry tailing of Rakha copper mine. 3 Biotech 7 (2), 132. Available at https://doi.org/10.1007/s13205-017-0757-y.

Harris, A.T., Naidoo, K., Nokes, J., Walker, T., Orton, F., 2009. Indicative assessment of the feasibility of Ni and Au phytomining in Australia. Journal of Cleaner Production 17 (2), 194–200. Available at https://doi.org/10.1016/j.jclepro.2008.04.011.

He, H., Ye, Z., Yang, D., Yan, J., Xiao, L., Zhong, T., et al., 2013. Characterization of endophytic Rahnella sp. JN6 from Polygonum pubescens and its potential in promoting growth and Cd, Pb, Zn uptake by Brassica napus. Chemosphere 90 (6), 1960–1965. Available at https://doi.org/10.1016/j.chemosphere.2012.10.057.

He, X., Xu, M., Wei, Q., Tang, M., Guan, L., Lou, L., et al., 2020. Promotion of growth and phytoextraction of cadmium and lead in Solanum nigrum L. mediated by plant-growth-promoting rhizobacteria. Ecotoxicology and Environmental Safety 205, 111333. Available at https://doi.org/10.1016/j.ecoenv.2020.111333.

He, L., Ren, Y., Zeng, W., Wu, X., Shen, L., Yu, R., et al., 2021. Deciphering the endophytic and rhizospheric microbial communities of a metallophyte commelina communis in different Cu-polluted soils. Microorganisms 9 (8), 1689. Available at https://doi.org/10.3390/microorganisms9081689, https://www.mdpi.com/2076-2607/9/8/1689/pdf.

Hołtra, A., Zamorska-Wojdyła, D., 2020. The pollution indices of trace elements in soils and plants close to the copper and zinc smelting works in Poland's Lower Silesia. Environmental Science and Pollution Research 27 (14), 16086–16099. Available at https://doi.org/10.1007/s11356-020-08072-0.

Hou, D., Lin, Z., Wang, R., Ge, J., Wei, S., Xie, R., et al., 2018. Cadmium exposure-Sedum alfredii planting interactions shape the bacterial community in the hyperaccumulator plant rhizosphere. Applied and Environmental Microbiology 84 (12), e02797-17. Available at https://doi.org/10.1128/AEM.02797-17, http://aem.asm.org/content/84/12/e02797-17.full.pdf#page=1&view=FitH.

Hou, D., Wang, K., Liu, T., Wang, H., Lin, Z., Qian, J., et al., 2017. Unique rhizosphere micro-characteristics facilitate phytoextraction of multiple metals in soil by the hyperaccumulating plant Sedum alfredii. Environmental Science & Technology 51 (10), 5675–5684. Available at https://doi.org/10.1021/acs.est.6b06531.

Hrynkiewicz, K., Dabrowska, G., Baum, C., Niedojadlo, K., Leinweber, P., 2012. Interactive and single effects of Ectomycorrhiza formation and Bacillus cereus on metallothionein MT1 expression and phytoextraction of Cd and Zn by willows. Water, Air, & Soil Pollution 223 (3), 957–968. Available at https://doi.org/10.1007/s11270-011-0915-5.

Hu, L., Robert, C.A.M., Cadot, S., Zhang, X., Ye, M., Li, B., et al., 2018. Root exudate metabolites drive plant-soil feedbacks on growth and defense by shaping the rhizosphere microbiota. Nature Communications 9 (1), 2738. Available at https://doi.org/10.1038/s41467-018-05122-7, http://www.nature.com/ncomms/index.html.

Idris, R., Trifonova, R., Puschenreiter, M., Wenzel, W.W., Sessitsch, A., 2004. Bacterial communities associated with flowering plants of the Ni hyperaccumulator Thlaspi goesingense. Applied and Environmental Microbiology 70 (5), 2667–2677. Available at https://doi.org/10.1128/AEM.70.5.2667-2677.2004.

Jain, S., Muneer, S., Guerriero, G., Liu, S., Vishwakarma, K., Chauhan, D.K., et al., 2018. Tracing the role of plant proteins in the response to metal toxicity: a comprehensive review. Plant Signaling and Behavior 13 (9), e1507401. Available at https://doi.org/10.1080/15592324.2018.1507401, http://www.tandfonline.com/loi/kpsb20.

Jiang, Y., Lei, M., Duan, L., Longhurst, P., 2015. Integrating phytoremediation with biomass valorisation and critical element recovery: A UK contaminated land perspective. Biomass and Bioenergy 83, 328–339. Available at https://doi.org/10.1016/j.biombioe.2015.10.013.

Jing, Y.X., Yan, J.L., He, H.D., Yang, D.J., Xiao, L., Zhong, T., et al., 2014. Characterization of bacteria in the rhizosphere soils of *Polygonum pubescens* and their potential in promoting growth and Cd, Pb, Zn uptake by *Brassica napus*. International Journal of Phytoremediation 16 (4), 321–333. Available at https://doi.org/10.1080/15226514.2013.773283.

Kicińska, A., Pomykała, R., Izquierdo-Diaz, M., 2022. Changes in soil pH and mobility of heavy metals in contaminated soils. European Journal of Soil Science 73 (1), e13203. Available at https://doi.org/10.1111/ejss.13203, http://onlinelibrary.wiley.com/journal/10.1111/(ISSN)1365-2389.

Kidd, P.S., Bani, A., Benizri, E., Gonnelli, C., Hazotte, C., Kisser, J., et al., 2018. Developing sustainable agromining systems in agricultural ultramafic soils for nickel recovery. Frontiers in Environmental Science 6, 369383. Available at https://doi.org/10.3389/fenvs.2018.00044, https://www.frontiersin.org/articles/10.3389/fenvs.2018.00044/full.

Kong, Z., Glick, B.R., 2017. The role of plant growth-promoting bacteria in metal phytoremediation. Advances in Microbial Physiology 71, 97–132. Available at https://doi.org/10.1016/bs.ampbs.2017.04.001.

Krzeszowska, E., Kokowska-Pawłowska, M., Krzeszowski, 2022. Distribution of selected critical elements in the carboniferous coal-bearing series of the Upper Silesian and Lublin Coal Basins (Poland). Acta Geologica Sinica (English Edition) 96 (1), 273–292. Available at https://doi.org/10.1111/1755-6724.14811, http://onlinelibrary.wiley.com/journal/10.1111/(ISSN)1755-6724.

Lapie, C., Leglize, P., Paris, C., Buisson, T., Sterckeman, T., 2019. Profiling of main metabolites in root exudates and mucilage collected from maize submitted to cadmium stress. Environmental Science and Pollution Research 26 (17), 17520–17534. Available at https://doi.org/10.1007/s11356-019-05168-0.

Leuschner, C., Ellenberg, H., 2017. Ecology of Central European Non-forest Vegetation: coastal to Alpine, Natural to Man-Made Habitats. Vegetation Ecology of Central Europe, vol. II. pp. 1–1093. Available at https://doi.org/10.1007/978-3-319-43048-5, http://www.springer.com/in/book/9783319430485.

Li, X., Sun, M., Zhang, L., Finlay, R.D., Liu, R., Lian, B., 2022. Widespread bacterial responses and their mechanism of bacterial metallogenic detoxification under high concentrations of heavy metals. Ecotoxicology and Environmental Safety 246, 114193. Available at https://doi.org/10.1016/j.ecoenv.2022.114193.

Li, T., Xu, Z., Han, X., Yang, X., Sparks, D.L., 2012. Characterization of dissolved organic matter in the rhizosphere of hyperaccumulator Sedum alfredii and its effect on the mobility of zinc. Chemosphere 88 (5), 570–576. Available at https://doi.org/10.1016/j.chemosphere.2012.03.031, www.elsevier.com/locate/chemosphere.

Liang, J., Shohag, M.J.I., Yang, X., Tian, S., Zhang, Y., Feng, Y., et al., 2014. Role of sulfur assimilation pathway in cadmium hyperaccumulation by Sedum alfredii Hance. Ecotoxicology and Environmental Safety 100 (1), 159–165. Available at https://doi.org/10.1016/j.ecoenv.2013.10.026.

Lodewyckx, C., Mergeay, M., Vangronsveld, J., Clijsters, H., Van Der Lelie, D., 2002. Isolation, characterization, and identification of bacteria associated with the zinc hyperaccumulator *Thlaspi caerulescens* subsp. Calaminaria. International Journal of Phytoremediation 4 (2), 101–115. Available at https://doi.org/10.1080/15226510208500076.

Lu, M., Huang, L., Wang, Q., Cao, X., Lin, Q., He, Z., et al., 2023. Soil properties drive the bacterial community to cadmium contamination in the rhizosphere of two contrasting wheat (Triticum aestivum L.) genotypes. Journal of Environmental Sciences 128, 117–128. Available at https://doi.org/10.1016/j.jes.2022.07.028.

Ma, Y., Oliveira, R.S., Nai, F., Rajkumar, M., Luo, Y., Rocha, I., et al., 2015. The hyperaccumulator Sedum plumbizincicola harbors metal-resistant endophytic bacteria that improve its phytoextraction capacity in multi-metal contaminated soil. Journal of Environmental Management 156, 62–69. Available at https://doi.org/10.1016/j.jenvman.2015.03.024, https://www.sciencedirect.com/journal/journal-of-environmental-management.

Ma, Y., Rajkumar, M., Freitas, H., 2009. Improvement of plant growth and nickel uptake by nickel resistant-plant-growth promoting bacteria. Journal of Hazardous Materials 166 (2–3), 1154–1161. Available at https://doi.org/10.1016/j.jhazmat.2008.12.018.

Ma, Y., Rajkumar, M., Moreno, A., Zhang, C., Freitas, H., 2017. Serpentine endophytic bacterium Pseudomonas azotoformans ASS1 accelerates phytoremediation of soil metals under drought stress. Chemosphere 185, 75–85. Available at https://doi.org/10.1016/j.chemosphere.2017.06.135.

Mahar, A., Wang, P., Ali, A., Awasthi, M.K., Lahori, A.H., Wang, Q., et al., 2016. Challenges and opportunities in the phytoremediation of heavy metals contaminated soils: a review. Ecotoxicology and Environmental Safety 126, 111–121. Available at https://doi.org/10.1016/j.ecoenv.2015.12.023, http://www.elsevier.com/inca/publications/store/6/2/2/8/1/9/index.htt.

Maluckov, B.S., 2015. Bioassisted phytomining of gold. Minerals, Metals and Materials Society, Serbia JOM 67 (5), 1075–1078. Available at https://doi.org/10.1007/s11837-015-1329-4, http://www.springer.com/materials/journal/11837.

McGrath, S.P., Zhao, F.J., 2003. Phytoextraction of metals and metalloids from contaminated soils. Current Opinion in Biotechnology 14 (3), 277–282. Available at https://doi.org/10.1016/S0958-1669(03)00060-0, http://www.elsevier.com/locate/copbio.

Meers, E., Vandecasteele, B., Ruttens, A., Vangronsveld, J., Tack, F.M.G., 2007. Potential of five willow species (Salix spp.) for phytoextraction of heavy metals. Environmental and Experimental Botany 60 (1), 57–68. Available at https://doi.org/10.1016/j.envexpbot.2006.06.008.

Meighan, M.M., Fenus, T., Karey, E., MacNeil, J., 2011. The impact of EDTA on the rate of accumulation and root/shoot partitioning of cadmium in mature dwarf sunflowers. Chemosphere 83 (11), 1539–1545. Available at https://doi.org/10.1016/j.chemosphere.2011.01.035, www.elsevier.com/locate/chemosphere.

Moreira, H., Pereira, S.I.A., Marques, A.P.G.C., Rangel, A.O.S.S., Castro, P.M.L., 2016. Selection of metal resistant plant growth promoting rhizobacteria for the growth and metal accumulation of energy maize in a mine soil: effect of the inoculum size. Geoderma 278, 1–11. Available at https://doi.org/10.1016/j.geoderma.2016.05.003, https://www.elsevier.com/inca/publications/store/5/0/3/3/3/2.

Pan, F., Meng, Q., Luo, S., Shen, J., Chen, B., Khan, K.Y., et al., 2017. Enhanced Cd extraction of oilseed rape (Brassica napus) by plant growth-promoting bacteria isolated from Cd hyperaccumulator Sedum alfredii Hance. International Journal of Phytoremediation 19 (3), 281–289. Available at https://doi.org/10.1080/15226514.2016.1225280, http://www.tandf.co.uk/journals/titles/15226514.asp.

Pongrac, P., Vogel-Mikuš, K., Regvar, M., Tolrà, R., Poschenrieder, C., Barceló, J., 2008. Glucosinolate profiles change during the life cycle and Mycorrhizal colonization in a Cd/Zn hyperaccumulator *Thlaspi praecox* (Brassicaceae). Journal of Chemical Ecology 34 (8), 1038–1044. Available at https://doi.org/10.1007/s10886-008-9502-7.

Prasad, M.N.V., 2003. Phytoremediation of metal-polluted ecosystems: hype for commercialization. Russian Journal of Plant Physiology 50 (5), 686–701. Available at https://doi.org/10.1023/a:1025604627496.

Qiu, Z., Tan, H., Zhou, S., Cao, L., 2014. Enhanced phytoremediation of toxic metals by inoculating endophytic Enterobacter sp. CBSB1 expressing bifunctional glutathione synthase. Journal of Hazardous Materials 267, 17–20. Available at https://doi.org/10.1016/j.jhazmat.2013.12.043.

Rajakaruna, N., Boyd, R.S., 2008. Edaphic factor. In: Jørgensen, S.E., Fath, B.D. (Eds.), Encyclopedia of Ecology. Academic Press, pp. 1201–1207. ISBN 9780080454054. Available at https://doi.org/10.1016/B978-008045405-4.00484-5.

Rajkumar, M., Prasad, M.N.V., Freitas, H., Ae, N., 2009. Biotechnological applications of serpentine soil bacteria for phytoremediation of trace metals. Critical Reviews in Biotechnology 29 (2), 120–130. Available at https://doi.org/10.1080/07388550902913772.

Reeves, R.D., Baker, A.J.M., Jaffré, T., Erskine, P.D., Echevarria, G., van der Ent, A., 2017. A global database for plants that hyperaccumulate metal and metalloid trace elements. New Phytologist 218 (2), 407–411. Available at https://doi.org/10.1111/nph.14907.

Salinitro, M., Montanari, S., Simoni, A., Ciavatta, C., Tassoni, A., 2022. Phytoextraction of arsenic, nickel, selenium and zinc from sewage sludge: from laboratory to pilot scale. Plant and Soil 481 (1–2), 195–212. Available at https://doi.org/10.1007/s11104-022-05630-y.

Sarkar, B., Mukhopadhyay, R., Ramanayaka, S., Bolan, N., Ok, Y.S., 2021. The role of soils in the disposition, sequestration and decontamination of environmental contaminants. Philosophical Transactions of the Royal Society B: Biological Sciences 376 (1834), 20200177. Available at https://doi.org/10.1098/rstb.2020.0177.

Seshadri, B., Bolan, N.S., Naidu, R., 2015. Rhizosphere-induced heavy metal (loid) transformation in relation to bioavailability and remediation. Journal of Soil Science and Plant Nutrition 15 (2). Available at https://doi.org/10.4067/S0718-95162015005000043.

Sheng, X., Sun, L., Huang, Z., He, L., Zhang, W., Chen, Z., 2012. Promotion of growth and Cu accumulation of bio-energy crop (Zea mays) by bacteria: implications for energy plant biomass production and phytoremediation. Journal of Environmental Management 103, 58–64. Available at https://doi.org/10.1016/j.jenvman.2012.02.030.

Sheoran, V., Sheoran, A.S., Poonia, P., 2009. Phytomining: a review. Minerals Engineering 22 (12), 1007–1019. Available at https://doi.org/10.1016/j.mineng.2009.04.001.

Singh, S., Parihar, P., Singh, R., Singh, V.P., Prasad, S.M., 2016. Heavy metal tolerance in plants: role of transcriptomics, proteomics, metabolomics, and ionomics. Frontiers in Plant Science 6, 165395. Available at https://doi.org/10.3389/fpls.2015.01143, http://journal.frontiersin.org/article/10.3389/fpls.2015.01143/pdf.

Sun, G.L., Reynolds, E.E., Belcher, A.M., 2019a. Designing yeast as plant-like hyperaccumulators for heavy metals. Nature Communications 10 (1), 5080. Available at https://doi.org/10.1038/s41467-019-13093-6, https://www.nature.com/ncomms/index.html.

Sun, W., Xiong, Z., Chu, L., Li, W., Soares, M.A., White, J.F., et al., 2019b. Bacterial communities of three plant species from Pb-Zn contaminated sites and plant-growth promotional benefits of endophytic Microbacterium sp. (strain BXGe71). Journal of Hazardous Materials 370, 225–231. Available at https://doi.org/10.1016/j.jhazmat.2018.02.003, https://www.elsevier.com/locate/jhazmat.

Sun, Yb, Zhou, Qx, An, J., Liu, Wt, Liu, R., 2009. Chelator-enhanced phytoextraction of heavy metals from contaminated soil irrigated by industrial wastewater with the hyperaccumulator plant (Sedum alfredii Hance). Geoderma 150 (1–2), 106–112. Available at https://doi.org/10.1016/j.geoderma.2009.01.016.

Tang, L., Hamid, Y., Zehra, A., Sahito, Z.A., He, Z., Beri, W.T., et al., 2020. Fava bean intercropping with Sedum alfredii inoculated with endophytes enhances phytoremediation of cadmium and lead co-contaminated field. Environmental Pollution 265 (Part A), 114861. Available at https://doi.org/10.1016/j.envpol.2020.114861, https://www.journals.elsevier.com/environmental-pollution.

Thijs, S., Langill, T., Vangronsveld, J., 2017. The bacterial and fungal microbiota of hyperaccumulator plants: small organisms, large influence. Advances in Botanical Research 83, 43–86. Available at https://doi.org/10.1016/bs.abr.2016.12.003, http://www.elsevier.com/wps/find/bookdescription.cws_home/675801/description#description.

Tirry, N., Kouchou, A., El Omari, B., Ferioun, M., El Ghachtouli, N., 2021. Improved chromium tolerance of Medicago sativa by plant growth-promoting rhizobacteria (PGPR). Journal of Genetic Engineering and Biotechnology 19 (1), 149. Available at https://doi.org/10.1186/s43141-021-00254-8.

Tsednee, M., Yang, S.-C., Lee, D.-C., Yeh, K.-C., 2014. Root-secreted nicotianamine from Arabidopsis halleri facilitates zinc hypertolerance by regulating zinc bioavailability. Plant Physiology 166 (2), 839–852. Available at https://doi.org/10.1104/pp.114.241224.

Ullah, I., Mateen, A., Ahmad, M.A., Munir, I., Iqbal, A., Alghamdi, K.M.S., et al., 2022. Heavy metal ATPase genes (HMAs) expression induced by endophytic bacteria, "AI001, and AI002" mediate cadmium translocation and phytoremediation. Environmental Pollution 293, 118508. Available at https://doi.org/10.1016/j.envpol.2021.118508, https://www.journals.elsevier.com/environmental-pollution.

Ustiatik, R., Nuraini, Y., Suharjono, S., Jeyakumar, P., Anderson, C.W.N., Handayanto, E., 2022. Mercury resistance and plant growth promoting traits of endophytic bacteria isolated from mercury-contaminated soil. Bioremediation Journal 26 (3), 208–227. Available at https://doi.org/10.1080/10889868.2021.1973950, www.tandf.co.uk/journals/titles/10889868.asp.

van der Ent, A., Baker, A.J.M., Reeves, R.D., Chaney, R.L., Anderson, C.W.N., Meech, J.A., et al., 2015. Agromining: farming for metals in the future? Environmental Science and Technology 49 (8), 4773–4780. Available at https://doi.org/10.1021/es506031u, http://pubs.acs.org/journal/esthag.

van der Ent, A., Echevarria, G., Erskine, P., 2016. Recent advances in developing tropical nickel agromining. In: Proceedings of the IMPC 2016, 28th International Mineral Processing Congress. Canadian Institute of Mining, Metallurgy and Petroleum, Australia.

Visioli, G., D'egidio, S., Sanangelantoni, A.M., 2015a. The bacterial rhizobiome of hyperaccumulators: future perspectives based on omics analysis and advanced microscopy. Frontiers in Plant Science 5, 752. Available at https://doi.org/10.3389/fpls.2014.00752, http://journal.frontiersin.org/Journal/10.3389/fpls.2014.00752/pdf.

Visioli, G., Vamerali, T., Mattarozzi, M., Dramis, L., Sanangelantoni, A.M., 2015b. Combined endophytic inoculants enhance nickel phytoextraction from serpentine soil in the hyperaccumulator Noccaea caerulescens. Frontiers in Plant Science 6. Available at https://doi.org/10.3389/fpls.2015.00638, http://journal.frontiersin.org/article/10.3389/fpls.2015.00638/pdf.

Vithanage, M., Kumarathilaka, P., Oze, C., Karunatilake, S., Seneviratne, M., Hseu, Z.-Y., et al., 2019. Occurrence and cycling of trace elements in ultramafic soils and their impacts on human health: a critical review. Environment International 131, 104974. Available at https://doi.org/10.1016/j.envint.2019.104974.

Wang, L., Lin, H., Dong, Y., He, Y., Liu, C., 2018. Isolation of vanadium-resistance endophytic bacterium PRE01 from Pteris vittata in stone coal smelting district and characterization for potential use in phytoremediation. Journal of Hazardous Materials 341, 1–9. Available at https://doi.org/10.1016/j.jhazmat.2017.07.036.

Wangeline, A.L., 2007. Fungi from seleniferous habitats and the relationship of selenium to fungal oxidative stress. PhD dissertation. Colorado State University.

Wei, J.L., Lai, H.Y., Chen, Z.S., 2012. Chelator effects on bioconcentration and translocation of cadmium by hyperaccumulators, Tagetes patula and Impatiens walleriana. Ecotoxicology and Environmental Safety 84, 173–178. Available at https://doi.org/10.1016/j.ecoenv.2012.07.004.

Wilson-Corral, V., Anderson, C., Rodriguez-Lopez, M., Arenas-Vargas, M., Lopez-Perez, J., 2011. Phytoextraction of gold and copper from mine tailings with Helianthus annuus L. and Kalanchoe serrata L. Minerals Engineering 24 (13), 1488–1494. Available at https://doi.org/10.1016/j.mineng.2011.07.014.

Wu, K., Luo, J., Li, J., An, Q., Yang, X., Liang, Y., et al., 2018. Endophytic bacterium Buttiauxella sp. SaSR13 improves plant growth and cadmium accumulation of hyperaccumulator Sedum alfredii. Environmental Science and Pollution Research 25 (22), 21844–21854. Available at https://doi.org/10.1007/s11356-018-2322-6.

Xu, C., Chen, X., Duan, D., Peng, C., Le, T., Shi, J., 2015. Effect of heavy-metal-resistant bacteria on enhanced metal uptake and translocation of the Cu-tolerant plant, Elsholtzia splendens. Environmental Science and Pollution Research 22 (7), 5070–5081. Available at https://doi.org/10.1007/s11356-014-3931-3.

Yan, A., Wang, Y., Tan, S.N., Mohd Yusof, M.L., Ghosh, S., Chen, Z., 2020. Phytoremediation: a promising approach for revegetation of heavy metal-polluted land. Frontiers in Plant Science 11, 513099. Available at https://doi.org/10.3389/fpls.2020.00359, https://www.frontiersin.org/journals/plant-science.

Yang, Q., Tu, S., Wang, G., Liao, X., Yan, X., 2012. Effectiveness of applying arsenate reducing bacteria to enhance arsenic removal from polluted soils by *Pteris vittata* L. International Journal of Phytoremediation 14 (1), 89–99. Available at https://doi.org/10.1080/15226510903567471.

Yuan, M., He, H., Xiao, L., Zhong, T., Liu, H., Li, S., Deng, P., Ye, Z., Jing, Y., 2014. Enhancement of Cd phytoextraction by two Amaranthus species with endophytic Rahnella sp. JN27. Chemosphere 103, 99–104. Available at https://doi.org/10.1016/j.chemosphere.2013.11.040.

Zaier, H., Ghnaya, T., Lakhdar, A., Baioui, R., Ghabriche, R., Mnasri, M., et al., 2010. Comparative study of Pb-phytoextraction potential in Sesuvium portulacastrum and Brassica juncea: tolerance and accumulation. Journal of Hazardous Materials 183 (1–3), 609–615. Available at https://doi.org/10.1016/j.jhazmat.2010.07.068.

Zemanová, V., Pavlíková, D., Hnilička, F., Pavlík, M., 2021. Arsenic toxicity-induced physiological and metabolic changes in the shoots of Pteris cretica and Spinacia oleracea. Plants 10 (10), 2009. Available at https://doi.org/10.3390/plants10102009.

Zhang, X., Kuzyakov, Y., Zang, H., Dippold, M.A., Shi, L., Spielvogel, S., et al., 2020. Rhizosphere hotspots: root hairs and warming control microbial efficiency, carbon utilization and energy production. Soil Biology and Biochemistry 148, 107872. Available at https://doi.org/10.1016/j.soilbio.2020.107872, https://www.elsevier.com/inca/publications/store/3/3/2.

CHAPTER SIX

Conserving metallophyte biodiversity

6.1 Introduction

Metallophytes or plants with evolved mechanisms to grow and thrive on metalliferous soils, possess pivotal values for research on their abilities of survival and habitation on minelands.

Natural selection over the years have driven the evolution of metallophytes. The plants, seeds, and propagules that could survive the selective pressure exerted by metal-rich substrates evolved as metallophytes. The mechanisms of metal tolerance vary considerably among plant species. Some species exhibit adapted internal mechanisms when growing on metalliferous soil, while their nonmetalliferous counterparts do not show these traits. However, others diverge completely, both morphologically and genetically, to form a new species. Not only they are preferable entities for research on physiological mechanisms, but are also optimal candidates for rehabilitation and restoration of postmine sites (Whiting et al., 2004). Since they demonstrate remarkable resilience and genetic adaptations, they are a part of varied ecological niches, making them a valuable resource for biodiversity conservation.

Mining operations influence biodiversity at several scales spatially, that is, site, landscape, regional, and global. The restricted geographical presence is, however, a key reason for their threat to extinction, since they are subjected to a wide range of anthropogenic activities as in mining operations, making their conservation extremely important.

Policymakers from various countries are agreeing on and focusing on the principles of sustainable development, a global maxim, and the framing of a global mining initiative is an important step in internal reform by the mining industry. Furthermore, the exploration, documentation, and conservation of these exceptional plants are indispensable to achieving sustainable development goals.

Metallophyte conservation involves both in situ and ex situ approaches. While in situ conservation involves establishment of protected areas within

Metallophytes in Mineral Mining
DOI: https://doi.org/10.1016/B978-0-443-30110-0.00007-6

mine site and conserving them, ex situ approach includes identifying metallophytes and preserving their germplasm, such as seeds or propagules. Additionally, metallophytes should be maintained in herbaria, germplasm banks, and botanical gardens at both regional and global scales, enabling their ex situ conservation (Baker and Whiting, 2008). These strategies not only play a crucial role in conserving genetic resources and protecting them from possible threats such as habitat loss and fragmentation, which are major causes of biodiversity loss and may lead to extinction.

Geobotanical exploration should be conducted at priority areas, specifically areas with significant mineral hotspots, naturally exposed metal outcrops and in the undisturbed site before development-using well-designed techniques to ensure that no species are lost due to mining operations, using well-designed techniques.

In the absence of a comprehensive database on metallophytes, several well-developed databases, such as Australian Metallophyte Spatial Database, METALS database, PHYTOREM database, and Global Hyperaccumulator Database can be used. These databases link information about native metal accumulating plants, habitat, and distribution across regions. Seven hundred twenty-one hyperaccumulator species have been listed in global hyperaccumulator database (Reeves et al., 2018), with many more yet to be identified.

6.2 The significance and potential value of metallophytes

Since metallophytes represent unique group of plants capable of tolerating and accumulating increased level of metals in their tissues and colonizing various metal-contaminated sites and metalliferous soils around the world, they are ecologically highly significant. In order to inhabit such extreme environments, metallophytes have developed specific adaptation mechanisms that enable them to thrive despite limitations caused by scarcity of nutrients, increased level of metal(loid)s and other nonoptimal soil characteristics. Research of metallophyte plant species can therefore offer valuable insights into developing of adaptation mechanisms and evolution of metal-stress tolerance and physiological processes on various types of metal-rich sites.

Metallophytes contribute to some of the key ecological services influencing the functional stability of ecosystems affected by heavy metal contamination or natural metal enrichment. They are prominent biodiversity resources that create or are important part of specialized vegetation communities adjusted to metal-rich sites. Some species are rare or endemics and adjusted only to particular locations, contributing to overall biodiversity of the area and sometimes even being important in conservation terms. Metallophytes can represent nurturing species for other organisms inhabiting such environments, providing food and shelter for diverse microorganisms and fauna. Their roots provide to the soil stabilization and erosion prevention while contributing to improvement of soil structure and improvement of water retention capacity. Additionally, metallophytes may reduce the soil metal bioavailability, lowering their mobility and impact on surrounding ecosystems or water bodies. In bare areas modified by anthropogenic activities resulting in metal pollution (metal mining, smelting, industrial activities), metallophytes often present pioneer species able to colonize such areas, giving the first signal that the biological revitalization is possible (Ranđelović et al., 2019). As such, they are initiating the process of ecological succession contributing to soil quality improvement and preparing the ground for secondary succession, as well as for further development and enrichment of plant communities (Ranđelović et al., 2014). Metallophytes are therefore important for the overall resilience of the ecosystems of metal-rich sites, as other plant species cannot survive and thrive in such extreme environments. Their growth can hamper the leaching and dispersion of metals from soil surface protecting the surrounding from contamination. Metallophytes take significant part in biogeochemical cycling of elements in the ecosystems of metal-rich sites. They can often regulate their flix within ecosystems by storing them in particular part of their tissues, making them less or more available to the surrounding environmental compartment and living systems. By producing organic matter, they contribute to the cycling of essential nutrients such as nitrogen, potassium, and phosphorous, improving overall fertility of soil during time. By absorbing carbon dioxide, they contribute to the carbon sequestration, which is particularly important for the ecosystems of extreme environments, as it makes them more resilient to the environmental stresses, including climate change (Kåresdotter et al., 2022).

Ability of metallophytes to inhabit metal-enriched sites has been the basis for their application in environmental technologies and biotechnologies. One of the key properties of metallophytes is their potential to be used in the remediation of contaminated sites, especially those affected by mining and other industrial activities generating inorganic pollution.

Phytoremediation, use of plants for cleaning up contaminated soils, completely relies on the utilization of metallophyte plant group for the accumulation or immobilization of present metals. In these terms, there is a growing need for geobotanists, environmental researchers, and mining companies to put in the efforts on screening metalliferous vegetation developed on mineralized areas, mining wastes, and other industrial sites in order to discover new biological resources that can be utilized in remediation of such legacy (Whiting et al., 2004; Baker and Whiting, 2008). Phytoremediation can offer more cost-effective, sustainable, and nature-based solutions for contaminated sites in comparison to some chemical or physical remediation techniques.

Few categories of metallophyte can be utilized in this sense: metal excluders, metal indicators, and metal (hyper)accumulators (Devi et al., 2020). While metal excluders offer immobilization of metals within their rhizosphere zone and consequent reduction of pollution dissemination, metal indicator species offer possibility to access the extent of contamination in certain area by means of plant tissue analysis. Metal accumulators are able to concentrate increased levels of metals in their tissues in comparison to average plant levels, whereas specific anomaly of strong accumulation of metals in plant shoots above certain determined threshold values is called hyperaccumulation. Metallophytes with hyperaccumulation abilities are recognized as the most effective to be used for phytoremediation purposes if they are able to produce adequate biomass yields. More than 700 plant species are recognized for their ability to hyperaccumulate metals, such as Ni, Zn, Cu, Co, and Mn (Reeves et al., 2018). Thanks to such unique ability within the plant world, hyperaccumulating metallophytes were utilized in technology for resource extraction called phytomining. Phytomining is a developing technology, recognized as environmentally-friendly alternative to traditional mineral resource extraction, aiming at concentrating increased levels of metals in plant above-ground tissues, later modified into material called bio-ore. Applying different technologies, it is possible to extract and recover targeted metal or metal compounds from the hyperaccumulator biomass, as well as to gain range of functional materials (Chai et al., 2022). Additionally, coupling bio-ore and bioenergy production from hyperaccumulators biomass represent an idea able to increase the overall feasibility of the phytoremediation process.

Another area of metallophytes application is food security and agroecosystem development. Due to their unique genetic traits that enable them colonization of metal-polluted sites, along with withstanding the unfavorable conditions such as drought or low fertility, metallophytes can be used as a unique genetic resource for the development of metal-tolerant agricultural crops via genetic engineering (Elango et al., 2022). Additionally, biofortification of agricultural crops for increased uptake of metals (especially those with recognized deficiency, such as Zn or Fe) by using metallophyte genetic material (Clemens, 2017) is a promising approach for improving food nutritional value and human diet.

In conclusion, the role of metallophytes in ecosystems of metal-rich sites is comprehensive. They contribute to various processes, including erosion prevention, biodiversity and genetic conservation, and promotion of ecological succession. Besides, they represent the most important plant group for ecological restoration of metal-degraded sites. Also, metallophytes represent outstanding natural resource for the development of various technologies such as phytoremediation, phytomining, or genetic crop improvement.

6.3 Guidelines and rules for metallophyte conservation

Excessive elemental concentrations in metalliferous sites lead to these areas being considered waste, and many metallophytes have disappeared in recent decades. However, some of these plants are rare, ecologically important, and often endemic to mining or metalliferous areas, which is why certain habitats, such as calaminarian grasslands, are considered areas of Special Areas of Conservation under Natura 2000 (Baker et al., 2010). Metallophyte conservation is a discipline of plant conservation that focuses on the protection of plants growing in sites with excessive concentrations of elements. It involves the application of conservation guidelines and rules established on the basis of scientific literature, international practices, and environmental policy.

Identification and mapping are the first steps in the conservation of plants, including those from metalliferous habitats. This includes systematic surveys to locate metallophyte species and characterize their populations (size and distribution) and habitat type (soil, geology, slope, climatic features, and vegetation characteristics). This would not only make it possible to prioritize rare endemic and relict species for conservation but also provide data on interaction with soil and other plants, pollination preferences, physiological and genetic mechanisms of metal tolerance, and hyperaccumulation. In addition, it would help to identify current and potential threats (mining, industrial pollution, soil degradation, and invasive species). Both in situ and ex situ measures can be taken. In situ protection refers to the preservation of existing habitats, the prevention of their further deterioration, and the implementation of measures to improve them. This includes various management plans at site level (control of mining activities, control of grazing, reduction of pressure from tourism, etc.), improvement of habitat characteristics through soil stabilization and erosion control, and reduction of habitat loss and species competition. Ex situ conservation includes activities outside the habitat itself, in laboratories, botanical gardens, nurseries, and so on, and aims to conserve the species themselves, their population and their genetic diversity. They start with the collection of seeds and cuttings, their conservation in seed banks, botanical gardens, and living collections, and development of protocols for germination, micropropagation and acclimatization. This also refers to tissue culture, which is very efficient in restoring populations. The reintroduction of rare and endangered species can efficiently restore the number of species in natural habitats. However, there are some measures that can help in this process, such as reducing competition between species, reducing soil disturbance and improving soil properties, mainly through soil amendments. In recent years, alumiunosilicate material has proven to be an effective method of improving soil and thus plant fitness (Djukić et al., 2025). The success of reintroduction can be assessed by measuring survival rates, reproduction, and integration into the community. In addition to their primary role in reintroduction, ex situ populations can also be used in scientific and educational practices.

The data obtained from metallophyte conservation activities can be implemented in practice into the legal recognition of metallophytes as species of conservation concern, and their inclusion in national biodiversity strategies and environmental legislation. This would also ensure that the surveys and conservation plan for metallophytes are included in future environmental impact assessments for mining projects and the development of land management guidelines specifically for metalliferous sites. In addition, the data can be integrated into open-access databases of species distribution, population trends, and conservation status for use by the scientific community, herbaria, botanical gardens, and conservation organizations.

6.4 Documentation of metallophyte diversity

The effective conservation of metallophytes depends to a large extent on accurate documentation of their diversity. Metalliferous habitats, although hostile to most species, harbor taxa that have evolved remarkable physiological and morphological adaptations to metal stress (Baker et al., 2010). Many of these species are stenoendemic, restricted to specific metalliferous outcrops, making them vulnerable to habitat loss and anthropogenic disturbance (Reeves, 2003).

The documentation of metallophyte diversity includes systematic floristic studies with accurate plant identification. The data obtained provides information on the habitat preferences of these species, and whether they can also grow on nonmetalliferous sites, thus identifying them as obligate or facultative metallophytes. Particularly important are the plants that are exclusively associated with metalliferous habitats, as they represent an irreplaceable part of the local diversity. Other factors that affect conservation measures include parameters such as the physico-chemical properties of the soil, the water regime, and the characteristics of the vegetation.

The importance of documenting metallophyte diversity has increased as many of these metallophyte species and their habitats have become increasingly threatened in recent decades by mining, industrial, and other activities, leading to land degradation and habitat loss, which is considered to be one of the main drivers of decline in biodiversity (Segan et al., 2016). Geographic information systems provide the necessary data on the distribution of metallophytes and their changes over time, allowing the identification of areas of conservation importance and represents the basis for conservation measures at national and international levels.

Apart from the importance for fundamental knowledge, the data on metallophyte diversity can be used not only for the planning and implementation of phytoremediation strategies but also for the use of plants in the sustainable extraction of trace elements, in a process known as phytomining. Of particular interest in these areas are hyperaccumulator plant species that are able to extract considerable elemental concentrations from the soil and accumulate them in their tissues (van der Ent et al., 2015). Documenting the biodiversity of metallophytes can therefore be used both for biodiversity conservation and for investigating the biotechnological potential of metallophytes and their use to improve environmental characteristics.

6.5 Conclusions

In essence, the degree of our understanding regarding metallophyte diversity, underlying survival mechanisms, and their implementation for eco-restoration is still limited. Systematic screening of plants present in metalliferous soils can help identify several hyperaccumulator species and priority candidates for conservation. Conservation strategies are pivotal in several aspects such as conserving germplasm of metallophytes for future research; protection of endemic and rare species against habitat loss; acting as a repository for research and application for eco-restoration projects; incorporation in plant breeding and crop improvement; and serving as an educational resource for spreading awareness regarding biodiversity conservation in minelands.

The ecological relevance of metallophytes with regard to biodiversity is one thing mining industry should be made aware of. This needs global-scale dissemination of data regarding metallophytes presence across mining sites. The amalgamation of academic interests with mining project makers focusing on restoration of mining site is mandatory for sustainable approach.

The answers to questions where to survey (priority areas with diverse and endemic metallophyte species), when to survey (before development so that no species loss has occurred), and how to survey (ecological methodologies) should be a compulsory part of mining plan.

References

Baker, A.J.M., Ernst, W.H.O., van der Ent, A., Malaisse, F., Ginocchio, R., 2010. Metallophytes: the unique biological resource, its ecology and conservational status in Europe, central Africa and Latin America. L.C. Batty and K.B. Hallberg (Eds.), Ecology of Industrial Pollution. Cambridge University Press (Ecological Reviews), Cambridge, pp. 7–40. https://doi.org/10.1017/cbo9780511805561.003.

Baker, A., Whiting, S., 2008. Metallophytes — a unique biodiversity and biotechnological resource in the care of the minerals industry. In: Fourie, A.B., Tibbett, M., Weiersbye, I.M., Dye, P.J. (Eds.), Proceedings of Third International Seminar on Mine Closure 2008, Australian Centre for Geomechanics, Perth, WA, Australia, pp. 13–20. https://doi.org/10.36487/ACG_repo/852_.

Chai, Y., Chen, A., Bai, M., Peng, L., Shao, J., Yuan, J., et al., 2022. Valorization of heavy metal contaminated biomass: recycling and expanding to functional materials. Journal of Cleaner Production 366, 132771. Available at https://doi.org/10.1016/j.jclepro.2022.132771.

Clemens, S., 2017. How metal hyperaccumulating plants can advance Zn biofortification. Plant and Soil 411 (1–2), 111–120. Available at https://doi.org/10.1007/s11104-016-2920-3.

Devi, P., Kumar, P., 2020. Concept and application of phytoremediation in the fight of heavy metal toxicity. Journal of Pharmaceutical Sciences and Research 12 (6), 795–804.

Djukić, D., Mirković, M., Andrejić, G., Aleksić, U., Mišljenović, T., Vujičić, M., et al., 2025. Green solution for lead pollution: phytoremediation with Festuca rubra and brushite-aluminosilicate geopolymer material. Plant Biosystems - An International Journal Dealing with all Aspects of Plant Biology 159 (3), 448–457. Available at https://doi.org/10.1080/11263504.2025.2485975.

Elango, D., Devi, K.D., Jeyabalakrishnan, H.K., Rajendran, K., Thoomatti Haridass, V.K., Dharmaraj, D., et al., 2022. Agronomic, breeding, and biotechnological interventions to mitigate heavy metal toxicity problems in agriculture. Journal of Agriculture and Food Research 10, 100374. Available at https://doi.org/10.1016/j.jafr.2022.100374.

Kåresdotter, E., Bergqvist, L., Flores-Carmenate, G., Haller, H., Jonsson, A., 2022. Modeling the carbon sequestration potential of multifunctional agroforestry-based phytoremediation (MAP) systems in Chinandega, Nicaragua. Sustainability 14 (9), 4932. Available at https://doi.org/10.3390/su14094932.

Ranđelović, D., Cvetković, V., Mihailović, N., Jovanović, S., 2014. Relation between edaphic factors and vegetation development on copper mine wastes: a case study from Bor (Serbia, SE Europe). Environmental Management 53 (4), 800–812. Available at https://doi.org/10.1007/s00267-014-0240-z.

Ranđelović, D., Mihailović, N., Jovanović, S., 2019. Potential of Equisetum ramosissimum Desf. for remediation of antimony flotation tailings: a case study. International Journal of Phytoremediation 21 (7), 707–713. Available at https://doi.org/10.1080/15226514.2018.1556590.

Reeves, R.D., 2003. Tropical hyperaccumulators of metals and their potential for phytoextraction. Plant and Soil 249 (1), 57–65. Available at https://doi.org/10.1023/A:1022572517197.

Reeves, R.D., Baker, A.J.M., Jaffré, T., Erskine, P.D., Echevarria, G., van der Ent, A., 2018. A global database for plants that hyperaccumulate metal and metalloid trace elements. New Phytologist 218 (2), 407–411. Available at https://doi.org/10.1111/nph.14907.

Segan, D.B., Murray, K.A., Watson, J.E.M., 2016. A global assessment of current and future biodiversity vulnerability to habitat loss-climate change interactions. Global Ecology and Conservation 5, 12–21. Available at https://doi.org/10.1016/j.gecco.2015.11.002.

van der Ent, A., Baker, A.J.M., Reeves, R.D., Chaney, R.L., Anderson, C.W.N., Meech, J.A., et al., 2015. Agromining: Farming for metals in the future? Environmental Science and Technology 49 (8), 4773–4780. Available at https://doi.org/10.1021/es506031u, http://pubs.acs.org/journal/esthag.

Whiting, S.N., Reeves, R.D., Richards, D., Johnson, M.S., Cooke, J.A., Malaisse, F., et al., 2004. Research priorities for conservation of metallophyte biodiversity and their potential for restoration and site remediation. Restoration Ecology 12 (1), 106–116. Available at https://doi.org/10.1111/j.1061-2971.2004.00367.x.

CHAPTER SEVEN

Case studies of valuable metals in agromining

7.1 Introduction

Rapid industrial and economic growth within the past few decades caused significant increase in use of raw mineral resources. Development of new and clean technologies, as well as the advancements of existing ones, increased the demand for specific elements that consequently evolved into critical or strategic resources. Some of the global critical raw materials are cobalt, copper, nickel, platinum group metals (PGMs) and rare earth elements (REEs) that have also been recognized as some of the strategic raw materials on the territory of European Union (The EU Critical Raw Materials Act, EC 2024/1252). The fact that the critical raw materials are connected to the supply risk or supply vulnerability has stipulated the need for their extraction from secondary sources or unconventional mineral resources. Thus, developing novel technologies and methods for recycling metals has emerged as a significant environmental, geopolitical, and economic issue.

Phytomining or agromining has emerged as an advanced approach in phytoremediation involving use of hyperaccumulating plants that are able to uptake and concentrate exceptional level of metals in their above-ground tissues. This process is suitable for establishment on various types of metalliferous soils and subeconomic/low-grade metal resources, such as ultramafic or laterite soils or mine wastes and tailing sites. Besides being a tool for environmentally-sound remediation of metal-enriched sites, hyperaccumulators are also regarded as potential secondary resources for metal recovery. Different site conditions, including soil physico-chemical properties, climate, and metal availability to plants can influence metal uptake rate and, consequently, successfull application of phytomining process. Therefore, in agromining systems, specific agrotechnological methods are developed not only to maximize the hyperaccumulator biomass production but also to maximize the plant metal uptake (Bani et al., 2015). Upon plant harvesting and adequate biomass treatment, metal or

Metallophytes in Mineral Mining
DOI: https://doi.org/10.1016/B978-0-443-30110-0.00008-8

its compound can be recovered in certain rate. Postharvest treatment and valorization of biomass remaining after phytoremediation and phytomining processes are subject of continual researches and upgrade (Chai et al., 2022). In case of phytomining, main biomass valorization product is metal that is afterward used in various industries (Kikis et al., 2024), whereas biomass originating from phytoextraction process usually has lower concentration of metals, and it is more often used in the production of energy, pharmaceutical or aromatic components, fibers, and so on (Pandey et al.,2023; Ranđelović and Pandey, 2023). The aim of agromining biomass valorization process is to develop easy, cost-effective procedure with minimal impact on the environment and the high metal recovery rate. The process should be adaptable to biomass variability (Kenney et al., 2013) and treatment of contaminated material (Li, et al., 2023).

Although the potential of plants for hyperaccumulation of metals was noticed in 1980s (Chaney, 1983), first phytomining field trials started in 1994 on soils containing subgrade levels of 0.35% of nickel in Nevada, USA, by using nickel hyperaccumulator *Streptanthus polygaloides* (Brooks et al., 1998). This study concluded that the economic feasibility of the process would be achieved with threshold of 1% nickel in plant dry biomass, giving biomass yield of 10 t ha^{-1}, resulting in nickel yield of 100 kg ha^{-1}. These studies were followed by experiment field trials on ultramafic soils of Tuscany, Italy, where phytomining experiment with nickel hyperaccumulator *Alyssum bertolonii* was conducted, producing 0.8% of nickel in plant dry matter with biomass yield of 9 t ha^{-1} and total nickel yield of 72 kg ha^{-1} (Robinson et al., 1997). Phytomining technology was finally patented by Chaney et al. (1998) as a method for the extraction of nickel, cobalt, and other metals from soil. Five-year long field trials (2005–2009) in temperate climate on ultramafic soils of Albania with nickel hyperaccumulator *Allysum murale* provided additional information on suitable agricultural technologies for enhancing yield and nickel phytoextraction, reaching biomass yield of 9 t ha^{-1} and total nickel extraction yield of 105 kg ha^{-1} (Bani et al., 2015). Phytomining trials on ultramafic soils in tropical climate conditions of Sabah, Malaysia, using hyperaccumulator *Phyllanthus rufuschaneyi* estimated potential nickel yield of 250 kg ha^{-1} (Nkrumah et al., 2019). Phytomining (agromining) technology proved to be economically viable technology in case of Ni (Chaney et al., 2018) while it is still in developing phase concerning some other valuable elements.

Phytomining technology cannot compete with conventional mining technologies in terms of efficiency and scope, yet it offers several advantages over traditional mining activities. Metalliferous soils or low-grade ores and mineral wastes, otherwise uneconomic for classical exploitation, can be subjected to agromining as unconventional but efficient metal extraction method. Environmental footprint of agromining is much smaller in comparison to the traditional mining; moreover, after the process of agromining, soil quality is enhanced and in certain cases, the area can become much more suitable for other agricultural activities. Some of the current drawbacks of phytomining include dependence of its general profitability from the metal pricing, limitation to the surface levels of soils or mine wastes, variability of the process efficiency depending on soil metal bioavailable pool, species extraction capacity, and obtained biomass yield. However, further development of biomass conversion and metal recovery methods should offer more promising solutions for the practical application of agromining in future.

7.2 Metal recovery from bio-ore to final products

In agromining process, hyperaccumulating plants produce biomass whose above-ground parts are enriched with specific metal(s) in concentrations above hyperaccumulation threshold (>0.01% dry weight for Cd, 0.03% dry weight for Co and Cu, 0.1% dry weight for Ni and Pb, 0.3% dry weight for Zn, and 1% dry weight in case of Mn, according to Reeves et al. (2018)). Diverse approaches, including genetic engineering and microbe assistance have been applied to increase metal uptake rate from metalliferous soils, followed by the development of technologies streaming for efficiency in metal recovery from secondary sources. Agromining-generated biomass is subjected to metal recovery process after the harvesting, most often using hydro- and pyrometallurgical processes. Hydrometallurgical processing consist of leaching the biomass with adequate leaching agent, where the metals precipitate in solution that is later purified for obtaining final product. Pyrometallurgical processing of biomass involves use of elevated temperatures where the obtained product (bio-ore) represents metal-enriched ash suitable for further treatment toward metal extraction. Most often, producing metal or its derivative such as salt or catalysts is investigated.

7.2.1 Nickel agromining

Nickel plays a significant role in modern technology, having broad range of applications such as for stainless steel production, alloy creation, battery production, or plating and coatings, representing a critical metal due to its role in transition toward clean energy. Nickel is considered among essential components for anticorrosion and rechargeable batteries, causing elevated consumption levels on global levels. Although it is considered fifth most abundant element in Earth's crust, world nickel demand has risen from 1.123 Mt in 2000 to Mt 2.385 in 2020, with an annual average growth rate of 3.8% (International Nickel Study Group, 2025). It has been estimated that primary sources of nickel will not be sufficient to fulfill the growing demand for the consumption of this element in the future, and low-grade Ni ores and secondary nickel resources came into spot as additional resources for obtaining nickel (Dilshara et al., 2024). Some of the recognized low-grade nickel sources are mine wastes, electronic batteries, and ultramafic and laterite soils. However, Ni-hyperaccumulating plants with >0.1% Ni in dry weight or 10%–20% of Ni in produced bio-ore are also prcieved as viable sources of nickel, exhibiting higher concentration than in some common nickel ores, and being free of silicates that can complicate the processing and direct extraction process from soils (van der Ent et al., 2015).

Nickel hyperaccumulation is observed in more than 500 plant species in the world, making it most abundant phenomena among metal hyperaccumulators (Reeves et al., 2018). Some of the discovered species are robust and can generate biomass quickly, making them suitable choice for application in agromining. Some Ni hyperaccumulators from Brassicaceae family in Balkans reached up to 3.49% of Ni in leaves and 2.15% of Ni in shoots (Jakovljević et al., 2022). Highest reported content of Ni in plant tissues (25.7% in dry mass) is recorded in latex of *Sebertia acuminata*, endemic tree species originating from New Caledonia (Jaffré et al., 1976). Agromining chain for nickel includes two main steps: (1) cultivation of hyperaccumulating plants, followed by the application of suitable agronomic technologies for generating high biomass yield while maintaining high Ni uptake, and (2) modification of harvested above-ground biomass into profitable Ni-containing products (Chaney et al., 2018). Both hydrometallurgical and pyrometallurgical methods were employed for recovering of nickel from biomass of various Ni hyperaccumulators. Besides recovery of elemental nickel, nickel salts and catalysts for organic reactions have also been produced from the treated biomass (Hazotte et al., 2020).

Barbaroux et al. (2012) tested two routes for nickel solubilization from hyperaccumulator *A. murale* (syn. *Odontarrhena muralis*). First route employed direct selective precipitation with sulfuric acid that proved ineffective due to the formation of strong complexes between nickel and organic matter in the solution, which prevented the precipitation of metal. Similar results were obtained with electroplating, extracting only 3.5% of Ni where no visible deposits on the cathode were found. The second route employed Cyanex 272 solvent extraction of nickel and electroplating process, achieving the Ni recovery efficiency of 72%. Following the similar route, Guilpain et al. (2018) employed water leaching batch and column experiments on biomass of *A. murale* and *Rinorea bengalensis*, using sodium hydroxide for selective precipitation. Ni extraction yield reached 90% for *A. murale*, followed by dissolved cations (K, Ca, and Mg) and dissolved organic compounds in leachate. But further precipitation of Ni by water leaching also turned out to be inefficient due to the formation and complexion with strong Ni–ligand bond, confirming the inefficiency of selective precipitation process for Ni recovery.

However, most of the researches concentrated on biomass combustion in order to increase the concentration of Ni in the bio-ore, and afterward develop the method for Ni recovery. Obtaining the adequate composition and structure of ash is keystone for further steps in the recovery process. Zhang et al. (2014) conducted researches to examine optimal conditions (temperature and duration) for the combustion of *A. murale*. Combustion temperature of 1100°C caused volatilization of Ni, while treatment at 550°C resulted in the production of bio-ore with up to 20% w.t. Ni. Besides Ni, obtained bio-ore contained elevated concentration of Ca, K, Mg, and Fe, requiring further purification treatments. Boominathan et al. (2004) studied characteristics of *A. bertolonii* and *Berkheya coddii* after combustion in furnace. Biomass of *A. bertolonii* contained up to 7.7% dry weight of Ni and produced bio-ore with up to 82% Ni, whereas biomass of *B. coddii* contained much lower initial concentration of Ni in biomass (0.49% dry weight), producing bio-ore with 8.6% of Ni. Additionally, bio-ore produced from *B. coddii* contained increased levels of Ca (up to 34%), which was identified as an obstacle for further Ni recovery process. These investigations pointed out that the problem of impurities and cation removal from bio-ore during Ni recovery process needs to be taken into account. Houzelot et al. (2018) studied kinetics, extraction rates, and yields of Ni, Mg, P, K, Fe, and Mn from *A. murale* ashes leached by sulfuric acid. Effects of steering speed, temperature, and sulfuric acid concentration were investigated to find optimal conditions for maximal Ni extraction, while minimizing the extraction of other elements. Results showed that the extraction of Ni was greatly influenced by the temperature and concentration of the sulfuric acid. For developed hydrometallurgy route of obtaining the Ni product, Houzelot et al. (2018) investigated the effects of upscaling on the characteristics of *A. murale* ashes, comparing those produced at the laboratory scale in a furnace and ashes produced at the pilot scale in a boiler. Although there were certain variation concerning elemental content and mass loss, they did not significantly influence the process of Ni recovery, enabling the optimization of this process at larger scale.

Further steps in technology were focused on obtaining different Ni products that would make the agromining process feasible, particularly those products that are of a high purity and can reach high prices on the market. Vaughan et al. (2017) demonstrated hydrometallurgical processing of nickel in a form of $Ni(OH)_2$ from ashes of tropical species *R. bengalensis* and *Phyllanthus securinegoides*. After the combustion, bio-ore material contained 5.5% and 12.7 wt.% of Ni, respectively. Prior to acid leaching, the ash was washed with water to remove the existing K_2CO_3. Sulfuric acid leaching treatment show Ni extraction rate of >90% at 60°C. Highest precipitation of nickel in the form of $Ni(OH)_2$ was exhibited using MgO in two stages, reaching >98% of precipitated Ni in the solution. The obtained $Ni(OH)_2$ contained impurities, estimated to reach 70% of the elemental Ni price on the market, requiring further purification steps or transformation into more profitable Ni compound. Barbaroux et al. (2012) developed a method for extracting Ni ammonium disulfate from *A. murale* ash. Nickel is extracted from bio-ore by leaching with 1.9 M sulfuric acid at 95°C, generating a leachate rich in Ni, Ca, K, P, Mg, and Fe. By adding sodium hydroxide, the pH of the leachate was raised to 5, and the leachate was further evaporated to reduce the volume and increase the Ni content. During the pH adjustment, Fe precipitated in the form of hydroxide, and had to be removed during the crystallization process. Cold crystallization with the addition of ammonium sulfate resulted in the formation of Ni ammonium disulfate, which was then subjected to further purification by resolubilization and recrystallization, leading to final 13.2% of Ni in produced salt. Mercier et al. (2012) designed and patented the technology for producing ammonium nickel sulfate hexahydrate (ANSH) salt from the bio-ore of *A. murale*. Zhang et al. (2016) managed to increase the purity of the obtained ANSH from the hyperaccumulator *A. murale* from initial 88.8% to 99%, by shortening the leaching duration, replacing the neutralization agent, and recrystallyzing the obtained ANSH salt to a higher purity. By optimizing this process, the overall reagent consumption has been reduced. Marković et al. (2021) applied the process defined by Zhang et al. (2016) on wild population of *A. murale* from ultramafic sites in Serbia containing approximately 0.33% of Ni in dry plant tissues. By employing the developed technology steps, bio-ore with 3.01% of Ni was generated. Acid leaching of ash was exhibited at 95°C using 2 M sulfuric acid, and purification steps were performed afterward to remove present Fe and Mg from a leachate. To neutralize Fe, a 10% suspension of calcium hydroxide was added, while sodium fluoride was added to remove Mg from the solution. The results showed that 44.4 g of bio-ore generated 5.2 g of ANSH of 73 % purity that can be further increased by adjusting the Mg removal from the leachate before ANSH crystalization. Afterward, Marković et al. (2025) applied additional purification step to obtain an ANSH salt of analytical purity (99.9%), generating 9.37 g of pure ANSH crystal from the initial 90 g of *A. murale* bio-ore (Fig. 7.1).

Biomass of nickel hyperaccumulators has also been investigated for the production of ecocatalysts, substances that accelerate chemical reactions without being transformed in it. Ecocatalysts showed higher selectivity and recyclability than traditional catalysts in many cases (Grison, 2021). During the procedure, above-ground parts of hyperaccumulating plants are calcinated at increased temperature (400°C–600°C) and treated with hydrochloric acid, generating catalyst after filtering the obtained solution.

Figure 7.1 Production of crystals of ammonium nickel sulfate hexahydrate salt of analytical purity from native populations of *Allysum murale* in Serbia to obtain crystals of analytical purity.

Grison et al. (2013) pointed out that chemical characteristics of *Psychotria douarrei* were suitable for the development of plant-based catalysts for multicomponent Biginelli reaction. Escande et al. (2015) further developed the Eco-Ni catalyst from *Geissois pruinosa* biomass, which successfully promoted chlorinated products by hydrochlorination of alkenes. Another ecocatalyst, Eco-Ni(0) was prepared by Escande et al. (2017) by utilizing leaves of *Psychotria gabriellae* for the transfer hydrogenation of diverse carbonyl compounds. These researches showed that biomass of hyperaccumulators can be additionally valorized within the sustainable and green chemistry approach.

Nickel recovery technologies from hyperaccumulator biomass are continually developing at the laboratory scale. However, certain processes of agromining chain are upscaled by the open-field pilot trials that have been conducted over the recent years. Long-term agromining trials have been conducted especially in Pojskë, Albania, with the aim of improving the agrononomy of *A. murale* for cultivation on Balkan ultramafic sites (Bani et al., 2015). Two European projects, Agronickel, developing Ni agromining on ultramafic land in Europe (2016–2019, https://project-wheel.faccejpi.net/agronickel/) and LIFE – AGROMINE: Cropping hyperaccumulator plants on nickel-rich soils and wastes for the green synthesis of pure nickel compounds (2016–2021, https://webgate.ec.europa.eu/life/publicWebsite/project/LIFE15-ENV-FR-000512/cropping-hyperaccumulator-plants-on-nickel-rich-soils-and-wastes-for-the-green-synthesis-of-pure-nickel-compounds), were dedicated to establish several pilot agromining trials on ultramafic sites in Albania, Austria, Greece, and Spain, encompassing various edaphic conditions and climate. These projects aimed to develop agromining systems for the utilization of soil metal resources in Europe and to examine if the phytomining could be the efficient and feasible ecological alternative to the metallurgical processes for metal recovery. Established field trials generated larger quantities of hyperaccumulator biomass that was later subjected to investigations in search for the optimal Ni recovery routes. By combining the various extraction procedures into a circular economy system, these projects demonstrated feasibility to acquire nickel-rich products from hyperaccumulator plants on a pilot scale. Agromining systems were also established in tropical conditions in Sabah, Malaysia, with the aim of increasing technology readiness level by demonstrating the feasibility of agromining in real-life conditions and set a field model for the succesfsull Ni agromining for *P. rufuschaneyi* (Nkrumah et al., 2019). In 2016, a spin-off company Econick of University of Lorraine, France, was established with the aim to produce metals from plants by using ecologically responsible approach. During 2023, Econick partnered with Aperam, globally known stainless steel–producing company, and formed joint venture Botanickel with the aim of thriving in sustainable production of nickel from biological sources for the industry of stainless steel. The additional goal of such efforts was to increase the technological readiness level and commercialize the agromining technology (van der Ent, 2023). Additionally, few start-ups were developed for implementing Ni agromining at the commercial scale in Albania, Greece, USA, and Malaysia (Rylott and van der Ent, 2025). In 2024, US Department of energy supported project PHYTOMINES - Plant Hyperaccumulators To MIne Nickel-Enriched Soils (https://arpa-e.energy.gov/sites/default/files/2025-01/PHYTOMINES_Project_Descriptions.pdf) of Metalplant phytomining start-up and Verinomics company that deals with plant genetic with the aim of developing sterile and high Ni-yielding hyperaccumulating crops suitable for the development of agromining in the USA.

Life cycle assessment, striving to evaluate sustainability and environmental impact of all stages of agromining chain was developed for the recovery of nickel from ultramafic soils in the Balkans with *A. murale* by producing ANSH salt (Rodrigues et al., 2016). In this process, erosion control was identified as one of the key parameters for sustainable Ni agromining, aiming to decrease initial Ni loss from soils. Additionally, heat generated from the biomass combustion process could be further utilized, reducing the overall costs of the process. In terms of land use, it has been concluded that agromining exhibits similar intensity of impact as conventional agriculture if implemented on the same areas. When applied on bare land, use of agromining crops is beneficial due to reduced toxicity and erosion, providing ecosystem services and eventually gaining profit.

Estimated average lifespan for profitable agromining, based on long-term researches and gathered data is 10–50 years (Bani et al., 2015; Nkrumah et al., 2016). Despite its inability to compete with traditional nickel mining methods in terms of profit and extraction rate, agromining is a valuable alternative or complementary technology that has a strong potential for recovering nickel from subeconomic ores that are not economically viable to extract by conventional mining methods. Moreover, environmental issues and large ecological footprint connected with conventional nickel mining are reccomending the search for the viable and more ecologically sustainable alternatives in suitable metalliferous areas.

7.2.2 Arsenic agromining

Arsenic is toxic metalloid element, 20th most abundant in nature. It is occurring in Earth's crust in all types of rocks in average concentration of 1.8 mg kg^{-1} (Kabata-Pendias, 2011). Arsenic is often associated with metallic and sulfide ores and organic shale. Besides its natural origin from rocks and volcanic eruptions, As can be deposited in environment due to the various anthropogenic activities, such as mining and smelting (particularly of copper or gold ores), through pesticides used for agricultural crops, or as pollutant generated during coal combustion. Presence of arsenic originating from natural sources in groundwater of some regions in the world is associated with potential risk for human health. Current uses of arsenic include pharmaceuticals, wood preservatives, agricultural chemicals, and applications in the mining, metallurgical, glass making, and semiconductor industries. Due to its application in clean technology transition, arsenic is defined as critical mineral (European Commission, 2023; US Geological Survey, 2022), and global market for arsenic is steadily growing.

Average content of arsenic in world soils is estimated to be 5 mg kg^{-1} (Punshon and Jackson, 2017), although in soils surrounding As mining areas, these values can reach even 40600 mg kg^{-1} (Karczewska et al., 2007). In aerobic soil conditions, arsenic is present as As^{+5}, while in anaerobic conditions, it exists as As^{+3}, showing increased mobility and toxicity than in previous oxidation state. Arsenic is greatly associated with clay, organic matter, and Fe–Mn oxides in soils (Kabata-Pendias, 2011; Ranđelović et al., 2020). Bioavailability of arsenic in soils depends on pH, oxidation state, and chemical form present. Arsenic acts as an analogue of phosphate, often entering the plants through phosphate transport system and disrupting the metabolic pathways in cells (Rehman et al., 2021). Average content of As in plants is less than 1 mg kg^{-1} (Abbas et al., 2018), whereas hyperaccumulation threshold is 1000 mg kg^{-1} (Reeves et al., 2018). Four species of ferns from genus *Pteris* (Pteris *vitatta*, Pteris *cretica*, Pteris *umbrosa*, and Pteris *longifolia*) have been determined as As hyperaccumulators, as well as the fern species *Pityrogramma calomelanos* (Meharg, 2003).

First discovered As hyperaccumulator was fern *P. vitatta*, which was found to uptake up to 2.26% As in its fronds dry mass (Ma et al., 2001). Several phytoextraction field sites were established in China for testing the efficiency of *P. vitatta* for As removal alone or with applied agrotechnical measures (Chen et al., 2018). With such phytoremediation technology applied, *P. vitatta* manage to remove 10%–17% of As from soil annually, while in intercropping system, soil and crops reached national quality standards in several years. Addition of phosphate fertilizers has been found to increase As uptake, as in contaminated soil of chemical plant in industrial area of Beijing, *P. vitatta* removed up to 155% (counting for approximately 13 kg ha^{-1}) higher level of As in comparison to the control (Wan et al., 2024).

Due to the high levels of accumulated arsenite in *P. vitatta*, the obtained above-ground biomass is considered a hazardous waste and requires proper disposal or adequate postharvesting treatment Yan et al. (2008) studied volatilization of As during *P. vitatta* incineration, where over 60% of As volatilized at temperatures below 400°C, and nearly quarter of total As concentration volatilized at 800°C. After incineration, the concentration of As got increased from 0.12% in dry biomass to 1.3% in the obtained bio-ore. Lei et al. (2019) manage to retain 76% of As in the solid residue by adding 10% of CaO during the combustion of *P. vittata* originating from As phytoremediation site, thereby reducing the presence of As in flue gas.

da Silva et al. (2018) investigated the possibility for 35% ethanol extraction of As from biomass of *P. vitatta*, coupled with magnesium for precipitation of soluble arsenic in form of Mg arsenate. This method resulted in total removal efficiency of 90% As from plant biomass, making it more suitable for safe deposition, while 10% of As remained in biomass leftovers. Another approach to As extraction from hyperaccumulator biomass was conducted by Jiang et al. (2022), who investigated the hydrothermal liquefaction of *P. vitatta* biomass by the addition of calcium oxide and calcium hypochlorite, resulting in 96.5% of As remaining in produced hydrochar. Aqueous phase from hydrothermal liquefaction treatments was afterward recycled as fertilizer for the promotion of *Allium sativum* growth, where the content of As in its edible parts was under limit values of national food standard, showing a promising route for practical utilization of As-hyperaccumulator biomass. Experiments with biomass of *P. cretica* were also performed in order to recover arsenic and obtain additional valuable products. Eze and Harvey (2018) studied the possibility for the extraction of accumulated As as Mg arsenate and its conversion to nanoparticles for medical use. Arsenic was extracted with ethanol–water with an efficiency of 94.3%, and then molybdic acid process was used for the recovery of arsenic and phosphorous, where majority (96% and 94%, respectively) was extracted as Mg-arsenate and Mg-phosphate, respectively. These compounds can be converted to more valuable forms, such as nanoparticles or fertilizers. Furthermore, As nanoparticles with potential use in medicine were obtained from Mg-arsenate using reduction with L-cysteine and sodium borohydride.

As arsenic has lower market value than other metallic components as Ni or Zn, so far agromining may represent only additional approach in the remediation of As-polluted sites. However, coupled with the increased rate of As recovery from hyperaccumulators biomass and the development of other valuable components for use in agriculture, clean technologies, medicine, or other industries, it could aid in financial viability of As-phytomining process in general.

7.2.3 Chromium agromining

Chrome is 24th most abundant element in nature, exhibiting toxic and carcinogenic effects to humans. It is naturally present in different rock types, with the highest average content of ~2000 mg kg^{-1} in ultramafic and ~200 mg kg^{-1} in mafic rocks (Yan et al., 2021). Chromium is extracted from chromite, accessory mineral of basic and ultrabasic igneous rocks. Ore deposits of chrome are formed by crystallization from magma, where South Africa and Zimbabwe are hosting some of the world's largest deposits. Main use of Cr is in the stainless steel production, for increasing its endurance and resistance to corrosion, while secondary uses are in refractories, pigments, ceramic glazes, and leather tanning. It has been considered critical mineral in the United States. Chromium is the important element in modern industry and so far irreplaceable in the production of stainless steel, so its market size is expected to grow in future.

Within the natural environments, chromium can be found in two main oxidation states: as Cr^{+3} and Cr^{+6}, with latter being more toxic. Average topsoil Cr content of world soils is 54 ppm (Kabata-Pendias, 2011). Its mobility and bioavailablilty to plants depend on its oxidation state, where Cr^{+6} has been much more mobile than Cr^{+3} form. Chromium is toxic to plants, causing oxidative stress, modifying the activity of various enzymes, and damaging plant cells. Additionally, it is much more mobile in roots than in shoots, resulting in mainly root accumulation (Sharma et al., 2023). However, several species are reported to reach Cr hyperaccumulation threshold (above 1000 mg kg^{-1}; Shahid et al., 2017), out of which *Leersia hexandra*, containing around 0.3% of Cr in the dry above-ground biomass, is identified as chromium hyperaccumulator in both field and laboratory conditions (Zhang et al., 2007). Older reference of Peterson and Girling (1981) reports 4.8% and 3% of Cr in leaves ash of *Sutera fodina* and *Dicoma niccolifera*, respectively. Additionally, Baker and Brooks (1989) report 0.24%

Cr in dry leaves of *S. fodina* and 0.15% of Cr in dry leaves of *D. niccolifera* from Zimbabwe. Due to a limited number of species that exhibit Cr hyperaccumulation trait, chromium agromining development is still an emerging option, while extraction technologies for the recovery of Cr from hyperaccumulator biomass are yet to be developed. However, due to the increased demand for chromium and its environmental toxicity, development of agromining options for chromium coupled with remediation of Cr-polluted or -enriched soils remains viable and perspective option for the future.

7.2.4 Zinc agromining

Zinc ranks as the 24th most prevalent element found in the Earth's crust and fifth most-used metal in the world (Chen et al., 2024). It is widely utilized in many industries, primary for steel galvanization to prevent rusting but also for electric batteries production, alloys with various metals, automotive industry, and as a catalyst in chemical industry. Additionally, compounds of zinc are applied in the production of rubber, glass, and paints and in pharmaceutical industry. Zinc has also been recently recognized as important metal for the transition to the clean energy technologies. Besides its role in industry, zinc is an essential element for living organisms taking part in vital metabolic processes. It is second most abundant trace element in human body, catalyzing activity of over 100 enzymes, regulating the expression of genes, and stabilizing the structure of proteins (Saper and Rash, 2009).

Primary sources of Zn are ores, particularly the mineral sphalerite (ZnS), while the secondary sources are mine tailings and wastes, smelter wastes, scraps, mineral dust, and metalliferous soils. Around 70% of the world produced Zn originates from primary sources, while the 30% comes from recycled Zn from secondary sources (Kaya et al., 2020). Global production of Zn has increased in 21st century, and its current annual production of around 13 Mt of metallic zinc (Kania and Saternus, 2023). Average Zn content in the Earth's crust and world soils is estimated to be around 70 mg kg^{-1} (Kabata-Pendias, 2011). Zinc is often geochemically associated with other elements such as Pb, Cu, and Cd in both minerals and soils (especially metalliferous calamine soils, rich in Cd, Pb and Zn; Fig. 7.2). Zinc and cadmium exhibit similar geochemical behavior, and Zn can be substituted in sphalerite by Cd. Similarly, Zn and Pb occur in sulfide minerals such as sphalerite and galena, sometimes being associated in mixed phases, whereas Zn and Cu can often be found linked in polymetallic

Figure 7.2 Metalliferous calamine soils rich in Zn, Pb, and Cd in former mining site of Plombières, Belgium.

deposits. In mining environments, these elements can also be found geochemically correlated within different environmental compartments (Elmayel et al., 2020; Jakovljević et al., 2020).

Total of 31 plant species have been recorded as Zn hyperaccumulator so far (Vaculík et al., 2024). Besides *Dichapetalum gelonioides*, tree species originating from nonmetalliferous soils of Borneo, other species are of herbaceous origin. While threshold value for Zn hyperaccumulation is set to be 3000 mg kg^{-1} in above-ground plant parts, average of metal concentration in the foliar tissues of Zn hyperaccumulators is recorded to be 18800 mg kg^{-1} (Xu et al., 2020). Some of the highest concentrations of Zn in above-ground tissues were reported in *Noccaea caerulescens*, with 4.4% in dry biomass on Pb/Zn mining site Pierrefitte in France (Reeves et al., 2001) and leaves of *Arabidopsis halleri* with 5.4% in dry biomass on nonmetalliferous site in Kowary, Poland (Stein et al., 2017). Due to the polymetallic nature of the initial metalliferous sites they developed on, some of the Zn hyperaccumulators also exhibit capacity to accumulate other elements, such as Cd and Pb. One of the prominent examples is *Thlaspi caerulescens*, well known for its ability to grow on soils contaminated with Zn, Cd, and Ni while posing the ability to accumulate significant amount of these elements in above-ground parts, often exceeding hyperaccumulation threshold (Milner and Kochian, 2008). Similar ability is found in *Sedum alfredii* originating from Pb/Zn mines in China, where concentration of metals in its above-ground parts reached up to 11116 mg kg^{-1} of Zn and 1090 mg kg^{-1} of Cd, exhibiting hyperaccumulation phenomenon for both elements. Likewise, concentrations of Zn and Cd in metallicolous populations of *A. halleri* from Eastern Europe reached up to 15 744 and 267 mg kg^{-1}, respectively (Bert et al., 2002). Multimetal uptake ability of some Zn hyperaccumulators needs to be taken into account in terms of postharvesting treatment of their biomass and the design of technologies for metal recovery.

Several methods of obtaining zinc from the biomass of hyperaccumulators have been investigated so far. Main utilized routes were pyrometallurgical and hydrometallurgical. Research focusing on the extraction of Zn from biomass has been primarily undertaken on *N. caerulescens* and *Sedum plumbizincicola* owing to their high Zn accumulation rates and capacity to produce substantial biomass. However, a particular technical approach for Zn recovery depends more on the final output than the plant species involved (Jovanović et al., 2022).

During pyrometallurgical processing, biomass is incinerated to produce Zn-rich bio-ores, which are later transformed to extract pure Zn. Combustion temperature influences Zn vaporization, as well as the vaporization of Cd and Pb, that are often additionally present in hyperaccumulator biomass. Therefore, the optimization of combustion temperature presents one of the most important steps for producing Zn-rich ashes. Zhong et al. (2015) studied characteristics of hyperaccumulator *S. plumbizincicola* biomass under thermal treatment with the aim of accessing the impact of different temperatures on the partitioning of Zn, Cd, and Pb in ash. Research showed that combustion temperatures lower than 650°C resulted in the accumulation of Zn (above 90%) and Pb (above 95%) in bottom ash, whereas increased temperatures lead to the volatilization of these elements, lowering their recovery rates. In the produced ashes, Zn could be found in pure and oxidized state, as well as in the association with carbonates and silicates. Additionally, optimal pyrolysis temperature was investigated for the production of Zn-rich biochar, revealing that in the temperature range of 450°C–750°C, >66% of Zn is preserved in biochar, offering a potential material for the development of the Zn recovery method (Zhong et al., 2016). Volatilizing temperature is generally lower for Cd than for Zn, thus enabling partial removal of Cd from the resulting combustion products. Influence of combustion treatments on the thermal degradation of *S. plumbizincicola* for the purpose of metal stabilization and proper disposal were examined by Zhang et al. (2021). Concentrations of Zn and Cd in final products increased from 350°C to 650°C in ashing and pyrolysis treatments, whereas their bioavailability was lower in produced biochar than in ash. Recovery of Zn was the highest at the lowest temperature, exhibiting 84.6% after ashing and 95.6% after pyrolysis at 350°C. Researches concerning issues of safe disposal and separation of metals in by-products have been additionally performed on the biomass *of S. alfredii* to access the behavior of Cd, Pb, and Zn under different temperature and gasifying agents (Cui et al., 2018). Gasification with N_2 and CO_2 was found to increase the volatilization rate of metals at higher temperatures (>700°C), stabilizing remaining metals in the accompanying gasification products. Process particularly designed to recover Zn and Cd from hyperaccumulator *N. caerulescens* (syn. *T. caerulescens*) was developed by Hazotte et al. (2017). Dry plants were incinerated at 620°C, temperature optimized for the preservation of metals in *N. caerulescens* bio-ore, achieving concentrations of 1.5% of Cd and 7.5% of Zn. Prior to the leaching procedure with sulfur acid, ash was treated with water to remove majority potassium ions. Separated leachate was filtered and Cd was extracted by cementation process with the Zn powder, reaching 68% of the total Cd content in the bio-ore. After that step, filtrate was subjected to further Zn extraction by selective precipitation. After dissolving Zn and the remaining Cd by leaching with nitrite acid, pH was increased to 13.5 to precipitate Cd in a form of hydroxide and then lowered to 11 to precipitate Zn hydroxide. More than 80% of Zn was initially recovered from the bio-ore, while approximately 65% of Zn was recovered from plant biomass. This research presents initial design to recover Zn and Cd from hyperaccumulators, and further investigations and purification steps are needed to produce chemicals suitable for use in industry.

Hydrothermal and hydrometallurgical routes were further explored for the recovery of value-added products from Zn hyperaccumulators biomass. Novel route of sustainable use of hyperaccumulators as nutrient-enriched hydrochar was recently proposed by Zhang et al. (2023). Biomass of *S. plumbizincicola* is transformed into hydrochar at 260°C, which was able to achieve good adsorption capacity for phosphate (52.5 mg g^{-1}) and ammonium (27.6 mg g^{-1}), showing potential to be applied as fertilizer in phytoremediation field trials. Similarly, Qian et al. (2018) showed that during subcritical hydrothermal liquefaction of *S. plumbizincicola* at 220°C, over 90% of Zn is released into solution, where it can be further recycled, while remaining biochar with low Zn content can be upgraded into porous carbon of high porosity and carbon dioxide capture potential. Biomass of *T. caerulescens* was used to make the Green Lucas reagent for the chlorination of primary, secondary, and tertiary alcohols by Losfeld et al.

(2012a, 2012b). After thermal treatment at 400 °C, the obtained ash was leached with hydrochloric acid, resulting in the formation of a solution rich in metal chlorides. Partial purification was then exhibited in order to gain a mixture of metals enriched with zinc chloride, which showed catalytic properties in the process of alcohol chlorination. Thus, the cleaner chemical methodology for obtaining $ZnCl_2$ in comparison to metallurgical technologies was obtained, proving the possibility of metal-enriched biomass use in organic synthesis. Losfeld et al. (2012a, 2012b) investigated the possibility of using *T. caerulescens* and *Anthyllis vulneraria* biomass for the production of Lewis acid catalysts. Using the cost-effective process of thermal treatment at 400°C and the treatment of ash with hydrochloric acid, the obtained mixture is concentrated and examined for use in various chemical reactions. The results confirmed the efficiency of synthesized catalysts in organic reactions such as Friedel–Crafts acylations, acetylation of anisole, and benzylation of toluene. Similarly, biomasses of *N. caerulescens* and *A. vulneraria* were sucesfully utilized for the preparation of Lewis acid catalysts and supported with K-10 montmorillonite, leading to the increased yield and selectivity in the Diels–Alder reaction. Escande et al. (2014) and Grison et al. (2015) managed to recover Zn from *A. vulneraria* in the form of Eco-Zn catalyst, proving its higher yields and decreasing of Sheldon's E-factor than conventional catalysts in the bromination of aromatic substrates. The process of preparation of plant-derived catalysts from calcinated biomass of metal-accumulating plants, including Zn hyperaccumulators, has been patented by Grison and Escarre (2012).

In recent years, green synthesis of metal nanoparticles from biomass that contains metals offered potential for novel routes of metal recovery. *Brassica juncea* metal-enriched biomass was used for the synthesization of carbon nanotubes and Cu/ZnO nanoparticles by precipitation of Cu/Zn chlorophyllin with NaOH and ethanol (Qu et al., 2012). Zinc oxide nanoparticles with photocatalytic activity against dichlorophenol were synthesized by using *S. alfredii* biomass from Zn chlorophyllin and NaOH (Wang et al., 2016).

The feasibility of Zn agromining was not examined in details so far, considering the fact that Zn products reach lower price on market in comparison to Ni and that Zn hyperaccumulation is often accompanied by enhanced accumulation of other metals, making purification technology more complex. Some limitation exists, as majority of so far known Zn hyperaccumulators are herbaceous plants of relatively low biomass, while intraspecific variability in Zn uptake rate is also found in many species

(Vaculík et al., 2024). Although there is a lack of Zn agromining examples conducted on larger areas, certain estimation of Zn yields have been made. Grison et al. (2015) calculated that *A. vulneraria* can remove up to 27 kg ha^{-1} of Zn in its harvestable biomass, whereas *N. caerulescens* can remove up to 17 kg ha^{-1} of Zn. Losfeld et al. (2012a, 2012b) estimated that *N. caerulescens* and *A. vulneraria* biomass production of 5 t ha^{-1} could yield approximately 60 kg ha^{-1} of Zn. Assisted accumulation practices and use of adequate agronomic technologies could potentially increase the Zn content in plants, influencing the overall gain of final products. Moreover, increased level of Zn pollution due to anthropogenic activities influences the total area of land that requires remediation measures concerning this element. However, the benefits of zinc agromining for environmental remediation, development of metal recovery technologies for biomass of hyperaccumulators, and their possible integration into multimetal extraction systems on sites polluted with more than one element make it a promising technology.

7.2.5 Manganese agromining

Manganese is 10th most abundant element in the Earth's crust, with an average content of ~770 mg kg^{-1} (Rudnick and Gao, 2003). Its common range in rocks is 350–2000 mg kg^{-1}, being highest in the igneous rocks (Kabata-Pendias, 2011). It is typically found in the form of manganese oxides or hydroxides, often associated with iron. Manganese components are used in different industries, such as steel production, production of batteries, aluminum alloys, and electronic components, and as a component of fertilizers and pigments. It is considered critical mineral due to its importance in global energy transition, and the increase demand for this element has led to requirement of new Mn sources (Harris and White, 2019).

Manganese in soils mainly originates from parent material and has a complex behavior depending on acidity, oxido-reduction processes, presence of organic matter, and so on (Gilkes and McKenzie, 1988). The mean content of manganese in world soils is estimated to be ~440 mg kg^{-1} (Kabata-Pendias, 2011). Manganese is essential element for plants, taking part in metabolic processes and photosynthesis; however, it can be toxic in elevated concentrations. It is readily taken by plants in the form of a bivalent cation. Average concentration of Mn in plant leaves ranges from 30 to 500 mg kg^{-1} of dry mass (Millaleo et al., 2010), whereas hyperaccumulation threshold is set to be 10000 mg kg^{-1}. Based on such criterion, total of 42 species are recorded as manganese hyperaccumulators in Global hyperaccumulator database (Reeves et al., 2018). Some of the most prominent species in this sense are *Virotia neurophylla* (dry leaf content of 5.5%), according to Jaffré (1979), *Gossia bidwillii* (dry foliar Mn content 2.6%), according to Fernando et al. (2006), *Phytolacca americana* (reaching Mn concentration in dry leaves up to 1.5%), according to Li et al. (2018), and *Phytolacca acinosa* (Mn concentration in dry leaves reaches up to 1.9%, according to Xue et al. (2004).

Manganese phytomining is at its very beginning and needs to be supported with additional field trials to prove its economic feasibility and adequate biomass yields. Moreover, recovery of Mn from hyperaccumulator biomass needs to be accessed via specifically designed technologies. Several Mn-hyperaccumulating plants from genus *Grevillea* were used for the ecological restoration of ultramafic sites in New Caledonia, being afterward transformed into specific Eco-Mn catalysts (Bihanic et al., 2020). After being treated with heat and HCl, polymetallic catalyst with Lewis and Brønsted acidity properties was produced, followed by the addition of the co/oxidant (oxygen or oxygen peroxide) for enhancing the diversity of the obtained ecocatalysts. Eco-Mn catalysts showed original polymetallic composition and flexible reactivity, being successfully used in a range of organic synthesis according to the green chemistry approaches. Other important step toward efficient recovery of Mn from biomass of hyperaccumulating *Zingiber officinale* was done by Oleszek et al. (2022). They develop feasible method for the separation of Mn from *Z. officinale* biomass, optimizing combustion temperature to 300°C and afterward extracting 90% of Mn with nitric acid after previously washing ash with dilute water to remove potassium impurities. These researches present initial steps toward better understanding of manganese recovery routes from hyperaccumulating plants, paving the way for wider application and acceptance of manganese agromining worldwide.

7.2.6 Rare earth elements agromining

REEs represent a group of 17 metallic elements in the periodic system, comprising of 15 elements of lanthanide series, with the addition of scandium and yttrium. Additionally, REEs can be divided into light REEs (from lanthanum to europium, including scandium) and heavy rare earth elements (HREEs, from gadolinium to lutetium, including yttrium). They occur naturally in Earth's crust and are relatively abundant; however, they are scarcely found in concentrated ore deposits accessible by traditional mining techniques. Economically significant deposits in that sense are limited to a certain parts of world. REE deposits are formed in primary deposits by magmatic processes or in secondary deposits by weathering and sedimentation. Average abundance of REEs in the Earth's crust is 170 mg kg^{-1}, with Ce and La being most abundant (63 and 31 mg kg^{-1}, respectively, according to Chen et al. (2024). Chemical characteristics of these elements makes them challenging for separation and purification, so current extraction and processing methods are marked by increased energy consumption, environmental burdens, and difficulties in achieving high purity of elements. REEs are important for modern industry, especially in production of high-tech devices, magnets, devices for energy preservation and renewal, in tools used for defense industry, and many more. Therefore, they are regarded as both strategically and critically important resources due to their essential role in diverse technologies. Due to their concentrated supply chains, they are considered to be geopolitically important raw material. Owing to current scarcity, various processes have been developed for their extraction from secondary sources, such as various electronic and industrial wastes, coal ash, and mine tailings (Swain and Mishra, 2019), leading to the diversification of the global market. Biotechnology methods for REE recovery using microorganisms and plants also offer a promising approach, potentially reducing dependence on traditional mining methods. Moreover, since REE tend to concentrate in the upper part of soil profile, potential for their phytoaccumulation is pronounced (Azizi et al., 2023). Soils enriched in REEs are often found within mining areas, roadsides, and fertilized agricultural areas (Tao et al., 2022).

Plants able to accumulate more than 1000 mg kg^{-1} of REEs in their above-ground parts are established as their hyperaccumulators, and so far 21 species with such ability have been recorded (van der Ent et al., 2021). Several attempts for recovery of REEs from hyperaccumulators biomass have been conducted to day, with fern *Dicranopteris linearis* (syn. Dicranopteris *dichotoma*) being applied in most studies as it naturally colonizes REE mining deposits. It has been noted that leaves of *D. dichotoma* contains 0.34%–0.7% of total REEs in its dry biomass (Xiao and Zhang, 2003; Xiao et al., 2003). Current extraction technologies developed for recovery of REEs from biomass can be assorted in the areas of pyrometallurgy, hydrometallurgy, and novel approaches. However, development of technologies for the recovery of REEs from biomass is still in its early stages. Attempts to recover REEs from biomass of *D. linearis* were conducted on dry and ashed plant biomass. Extraction of REEs by different extraction solutions from dried biomass of *D. linearis* has been investigated by Chour et al. (2020). They applied deionized water, citric acid, citrate, sulfur acid, and EDTA as potential extractants, where strong complexing agent and strong acid generated extraction yield of more than 80%. Precipitation of REEs with oxalic acid leads to the formation of oxalates with the avoidance of aluminum precipitation, while calcination step was applied to finally obtain REE oxides. Extraction yields for EDTA and H_2SO_4 were 87.5% and 72%, respectively. However, impurity profile was dominated by Na in case of H_2SO_4 and Ca in case of EDTA. Leaching with sulfuric acid resulted in lower concentration of Al in the final solid and lower overall costs of the process. Process for recovery of REEs from *D. dichotoma* based on enhanced ion-exchange leaching in HNO_3 by using an ion-exchange resin was initially developed by Chour et al. (2018). Residual solution and completion ions were afterward eliminated by using deionized water and then additionally eluated with HNO_3. Recovery yield of REEs for this process was 78%, while the purity of the obtained REEs was 81%. Additionally, organosolv biomass fractionation and fast pyrolysis process resulted in separation of lignin and phenols as value-added products, revealing potential for simultaneous REE recovery and value chemical generation. Jally et al. (2021a, 2021b) developed a method for recovering REEs from *D. linearis* by using combustion and leaching. Upon combustion at 550°C, ash containing 3% of REEs was generated. Due to the increased contents of Al and Si in *D. linearis*, alumosilicate matrix was formed that stabilized REEs, disabling effective acid leaching. Sodium hydroxide solution is therefore inserted as a pretreatment step being applied to dissolve Al–Si matrix and concentrate REEs. Gain solid was then rinsed with water to lower pH, reduce Al and Si, and increase REE content in solid. Remaining solid is leached with nitric acid for final mobilization and extraction of REEs. By using this approach, Jally et al. (2021a, 2021b) recovered 74% of REEs from its initial content in hyperaccumulator biomass. Guided by results of previous researches, selective extraction of REEs in combination with derivation of value-added products from biomass of *D. linearis* was studied by Xie et al. (2023). Sulfuric acid–assisted etanol fractionation was applied for the selective extraction of REEs, resulting in the retention of 94.5% of REEs and 87.5% of Ca in solid phase, whereas other cations migrated into liquid. Increased separation and purification of REEs from solid fraction was afterward obtained by using oxalic acid and ammonium sulfate, reaching recovery rate of 89% and high purity of 97%. On the other hand, Li and Zhang (2024) studied the process of transfer of REEs into biocrudes by treatment with different acids, revealing that almost total recovery of REEs was achieved with usage of sulfur acid during microwave-assisted hydrothermal carbonization. Within the obtained ranges of purity, it has been estimated that the yield of REEs from biomass of *D. linearis* (reaching biomass yield up to 15 t ha^{-1} and 0.2% of REEs in dry shoots) would be up to 30 kg ha^{-1} (Jally et al., 2021a, 2021b). Additionally, other hyperaccumulating species could also present suitable material for the REE recovery process. One such promising species is *P. americana*, a plant that is able to generate higher biomass yield and is recognized for its ability to hyperaccumulate HREEs, Al, and Mn in leaves (Liu et al., 2021). Content of REEs in its leaves can reach up to 0.1%, while some recent researches suggest that accumulating trait for REEs is a characteristic trait conserved among the several species of *Phytolacca* genus (Grosjean et al., 2019).

Generally, extraction pathways that have been developed for REE hyperaccumulator *D. linearis* so far have shown deficiencies in terms of separation and purification, representing challenges for economically feasible REE recovery from biomass and agromining process. More studies are needed for the optimization of REE phytoextraction and recovery processes and consequential creation of feasible agromining chain, with additional inclusion of other hyperaccumulating species more suitable for the recovery process.

7.2.7 Thallium agromining

Thallium is soft metal known for its toxicity, often hosted by sulfide minerals associated with hydrothermal mineralization or potassium minerals in igneous rocks. Thallium sulfate was used as rodenticide in the past, but due to the strong toxicity, it was banned for further use (Genchi et al., 2021). Nowadays, thallium is mostly used in electronic industry, glass production, and medicine. Antropogenic sources of thallium are coal combustion, smelting and refining of metal sulfides, and cement industry.

Average concentration of thallium in soils is up to 1 mg kg^{-1} (Xiao et al., 2004). It is often found in residual soil fraction as nonmobile; however, in contaminated or naturally Tl-enriched soils, it can pose significant risk for the environment, as it is readily available for uptake by biota (Vaněk et al., 2009). Median content of Tl in unpolluted European soil is 0.66 mg kg^{-1} (Salminen et al., 2005), whereas soils around thallium deposit in Lanmuchang, China, contain up to 124 mg kg^{-1} (Xiao et al., 2004) and in mining soils of Allchar deposit, North Macedonia, they reach up to 500 mg kg^{-1}. Average content of Tl in plants is 0.05 mg kg^{-1} (Kabata-Pendias, 2011), while hyperaccumulation threshold for this element was set up to 100 mg kg^{-1} (Reeves et al., 2018). However, so far only few plant species are recognized as Ti hyperaccumulators, including *Biscutela leviegata* (with above 3.2% in its leaves, according to Salinitro et al. (2024), *Iberis intermedia* (0.4% in above-ground biomass, according to LaCoste et al. (1999), and *Viola arsenica* (up to 5.9% Ti according to Jakovljević et al. (2023).

Processes for recovering Tl from hyperaccumulator biomass are in its very first stages. Grison et al. (2015) utilized leaves of *I. intermedia* to synthesize the Eco-Tl catalyst for selective protection/deprotection procedures, especially of dithioacetal deprotection. Other functional components, such as organo-metal complexes or nanoparticles of thallium could additionally be synthesized from hyperaccumulators biomass (Robinson and Anderson, 2021). However, certain estimation of economic viability of thallium agromining have been made, as Tl phytoextraction is considered profitable process due to the high costs of Tl on the world market (ranging US$300–US$3000 per kg). Leblanc et al. (1999) concluded that Tl hyperaccumulator biomas yield of 10 t ha^{-1} containing 0.08% of Tl in dry above-ground biomass produced by agromining on larger areas could be a profitable option. Even small-scale agromining with *I. intermedia* could reach profitable production with 0.1% of Tl in above-ground biomass and biomass yield of 10 t ha^{-1} (Corzo Remigio et al., 2022). To gain more precise insight into thallium agromining perspective research is needed in the selected natural or anthropogenically Tl-enriched soils.

7.2.8 Noble metal agromining

Noble metals represent a group of metallic elements (gold, silver, and PGMs) characterized by high resistance to oxidation and corrosion in various environments. Although they occur in different mineralization settings and magmatic, hydrothermal, or sedimentary deposits, their content in Earth's crust is considerably low. Noble metals are usually accumulated in plants in low contents due to their scarcity and low solubility in soils. Search for noble metal hyperaccumulators and accumulators in natural and induced conditions are ongoing tasks in the field of phytomining.

Gold has chalcophyllic properties and can be found in hydrothermal deposits; however, due to the resistance to weathering processes, it can be concentrated in placer deposits. Largest gold deposits are placed in Australia and Russia. Gold has been traditionally used for jewelry and as a store of value; however, in recent times, it also finds various application in electronics industry and medicine. Global market is expanding and demand for gold is expected to grow in future, so Au recovery from secondary sources is expected to develop toward creation of green technology routes and increased extraction efficiency. Average content of gold in rocks is 0.0005–0.005 mg kg^{-1} (Pitcairn, 2013), while in global soils, it reaches average content of 0.0006–0.02 mg kg^{-1} (Kabata-Pendias, 2011). In plants, average content of gold reaches 0.01 mg kg^{-1}, while hyperaccumulation threshold is set to 1 mg kg^{-1} in dry above-ground tissue (Anderson et al., 1998). Plants are able to accumulate gold in its soluble state; however, no natural hyperaccumulator of gold has been recorded so far, and the capacity of plants to accumulate this element has been tested through induced phytoaccumulation.

Anderson et al. (1998) tested induced accumulation of gold on *B. juncea* using ammonium thiocyanate, leading to the uptake of up to 19.4 mg kg^{-1} in above-ground plant parts. Economic feasibility calculations gave threshold value of 17 mg kg^{-1} of dry weight in *B. juncea*, which was exceeded in this experiment. Lamb et al. (2001b) applied cyanide for inducing gold accumulation in *B. juncea* and *Cichorium* spp., resulting in increased uptake of 326 mg kg^{-1} of gold in leaves of *B. juncea* and 164 mg kg^{-1} in total plant biomass of *Cichorium* sp. To test the phytomining profitability assumption in field conditions, Anderson et al. (2005) conducted first agromining field trials with *B. juncea* at Fazenda Brasileiro Au mine, Brazil, applying sodium cyanide irrigation. After lixiviant treatment, *B. juncea* accumulated up to 20% of Au present in the soil during one growing season, reaching maximum of 63 mg kg^{-1} in plant tissues at 0.6 mg kg^{-1} of gold ore present in the substrate. Although toxicity of ammonium thiocyanate is lower in comparison to traditionally used cyanide, process for gold mobilization has to be carefully controlled to prevent leaching and pollution (Anderson et al., 1999).

Rodriguez et al. (2007) discovered that desert woody plant *Chilopsis linearis* can accumulate up to 179 mg kg^{-1} Au in its leaves after 35 days of growth on gold-enriched medium. Gold in tissues was stored in the form of nanoparticles of radius up to 9.4 Å, which depended on the initial concentration of gold in the substrate. Such a phenomenon can be further applied in green synthesis and chemistry. For example, Sharma et al. (2007) proved catalytic activity of Au-nanoparticle-enriched tissues in reduction of 4-nitrophenol. Extraction of gold from combusted plant material has been studied by Lamb et al. (2001), who leached gold-enriched biomass ash with HCl. Aqueous solution was then treated with ascorbic acid, and filtered solid residue precipitated metallic gold upon heating up to 550°C. Similarly, Dinh et al. (2024) analyzed the formation of gold particles in ash of woody plants after three-step leaching. Upon the biomass combustion, bottom ash was leached with water, acid (HCl), and base (NaOH), producing 12.1 mg kg^{-1} of gold in the derived leaching residue, 242-fold greater than in its starting biomass content. Gold particles present in bottom ash prior to leaching initially reached purity of above 98%. Initial experiments concerning agromining of gold have been conducted proving that it could reach economic viability. However, applied technologies still need refining and innovation to improve environmentally safe enhancement of the extraction and high recovery rate. The demand for gold and for more sustainable mining technologies is rising, so agromining could become more commercial activity for obtaining gold from secondary resources in the future.

Silver is rare element, often found associated with lead, copper, and gold. During weathering process, it is released from parent materials, and its concentration in unpolluted soils is up to 1 mg kg (Kolesnikov et al., 2020). Average silver concentration in plants are reported to range from 0.03 to 2 mg kg^{-1} (Kabata-Pendias, 2011). Increased level of Ag in plants (5–10 mg kg^{-1}, according to Kabata-Pendias (2011)) result in the development of toxicity symptoms, including disruption of metabolic pathways, photosynthetic pigments and nutrient uptake (Dresler et al., 2019). Natural hyperaccumulation of Ag in vascular plants has not been confirmed to day, although certain macrofungal species from genus *Amanita* from uncontaminated sites were able to naturally accumulate silver, reaching up to 1235 mg kg^{-1} in Amanita *strobiliformis* (Borovička et al., 2007). Investigations of Sagiroglu et al. (2006) on the polluted soils of Pb-Zn Keban mining district in Turkey have shown that *Euphorbia macroclada, Astragalus gummifer*, and *Verbascum cheiranthifolium* can accumulate 0.53, 0.54, and 0.37 mg kg^{-1} of Ag in its above-ground parts, respectively, while enrichment factor was >1 in case of *E. macroclada*. Additionally, Haris and Bali (2007) have studied the efficient uptake of Ag from aqueous solution in *B. juncea* and *Medicago sativa*. At concentrations ranging from 500 to 10,000 mg kg^{-1} Ag-nitrate *B. juncea* accumulated up to 12.4 w.t.% and *M. sativa* up to 13.6 w.t.% of silver that was stored in plant tissues in the form of Ag nanoparticles, showing hyperaccumulation ability as well as the potential for nanoparticle production and utilization in green chemistry.

The PGMs are represented by six transitional metal elements: platinum, palladium, iridium, ruthenium, rhodium, and osmium. They are characterized by their strong catalytic activity, high melting points, and heat and corrosion resistance. PGMs are used for catalyzing various chemical reactions, in fuel-processing reactions, for the pollution emission control, in clean technologies products, in medical products, and so on. They are considered critical raw materials in Europe. PGMs are usually associated with magmatic or hydrothermal sulfide deposits. Average concentration of PGMs in bulk crust is 0.0015 mg kg^{-1}, while in the richest PGMs deposit in the world—Bushveld complex in South Africa—average concentrations reach up to 5670 mg kg^{-1} (Pilchin and Eppelbaum, 2017), whereas palladium is usually found in lithosphere in greater abundance than platinum (0.015 mg/kg and 0.005 mg/kg, respectively, according to Ravindra et al. (2004)).

Content of PGMs in various plants ranges from 0,00009 to 14.6 mg kg^{-1} as reviewed by Feichtmeier and Leopold (2014), being accumulated in limited amounts from soil to plant tissues and predominantly stored in plant roots. Nemutadani et al. (2006) tested Ni hyperaccumulator *B. coddii* for platinum and palladium content in metalliferous outcrops in South Africa. Results showed the average accumulation of platinum in leaves of 0.22 mg kg^{-1} and accumulation of palladium of 0.71 mg kg^{-1}, with concentration factors in reference to soil content of 5.5 and 10.1, respectively. This research revealed the accumulation ability of *B. coddii* for PGMs. (Odjegba et al., 2007) studied uptake and physiological effects of PGMs on *Lactuca sativa* in nutrient solution. After exposure to 1 mg kg^{-1} of PMGs, *L. sativa* accumulated 52.6 mg kg^{-1} Pt and 27.1 mg kg^{-1} Pd. Similarly, *Hordeum vulgare* was cultivated in hydroponics and exposed to 1–50 μm l^{-1} Pd derived from Pd nanoparticles by Battke et al. (2008)). Palladium uptake in leaves was 17 mg kg^{-1} for the highest concentration of Pb in aqueous solution formed from particles of larger diameter, which corresponded to the content of Pd in leaves of *H. vulgare* in hydroponic solution formed from smaller nanoparticles at concentration of 6 μm l^{-1}. Potential of *Chrysopogon zizanoides* for palladium phytoremediation was tested in water solution treated with 10–120 ppm Pd by Masinire et al. (2022). Removal efficiency was the highest (83%) at concentration of 10 ppm (8.33 ppm Pd) after 3 weeks of growing, whereas removal efficiency was the lowest (20%) at concentration of 120 ppm (24 ppm Pd). Concentration in leaves was much lower than in roots, reaching 0.04–0.15 ppm across the Pd solution gradients.

There has been limited number of attempts to extract PGMs from plants biomass. At 2012, University of York developed project PHYTO-CAT with the aim of inducing phytoaccumulation of palladium (among several other elements) in threshold concentration of 1000 mg kg^{-1} in plants grown on mine waste in order to develop catalysts (Robinson and Anderson, 2018). During this project, solution of NaCN was applied to *Cannabis sativa*, which was able to reach only 62.4 mg kg^{-1} of Pd in dry biomass due to enhanced accumulation and deterioration effects of other present elements in the mining soil. Parker et al. (2014) demonstrated catalytic activity in pyrolized biomass of *Arabidopsis thailiana* containing 0.5% mg kg^{-1} Pd. Grown *Arabidopsis* produced palladium nanoparticles that were proved to have catalytic behavior for variety of the Suzuki–-Miyaura reactions. Later on, Harumain et al. (2017) showed that the content of Pt in *Arabidopsis* grown in hydroponics with the addition of potassium tetrachloropalladate needs to reach 12000–18000 mg kg^{-1} in dry mass in order to be comparable to the standard commercial Pd catalysts. In 2023–2024, University of York started project PETAL (https://gtr.ukri.org/projects?ref=BB%2FX011232%2F1) aiming to increase PGMs uptake to plants by utilizing cyanogenic plants and bacteria instead of traditionally used chemicals and test the recovery of Pd in the form of catalysts from pyrolysed Pd-enriched biomass.

Noble metals are potential candidates for agromining due to their scarcity and possibility to be accumulated in some plant species via enhanced phytoaccumulation. Use of cyanide solution, often applied in traditional mining settings for the extraction of noble metals, can cause significant environmental damage, and cyanide-producing potential of some biota can be potentially employed as an environmentally-friendly substitute. Methods for recovery of noble metals from hyperaccumulator biomass are barely existing and further interdisciplinary researches coupled with real-world field trials are necessary in order to generate larger quantities of enriched biomass and investigate noble metals agromining potential in more detail.

7.3 Conclusion

Increase of metal pollution over the world, growing demand for raw materials and discovery of plant species able to accumulate extraordinary level of certain metals in their harvestable tissues all generated development of novel technology called phytomining or agromining. This technology aims to extract desirable metals from soil to strategically cultivated plants and then use adequate technologies to recover it into valuable metallic products. Development of agromining technology is currently at various stages for different elements, from initial trials (PGMs, chromium, REEs) to developed technologies used on commercial scale (case of Ni). Increased demand for critical or strategic raw materials calls for investigation of various methods and approaches for recovery of metals from hyperaccumulator biomass as valuable secondary resource. Besides obtaining a raw material, technologies for acquiring various functional materials have been developed by metallurgical or chemical methods. Metal salts (with Ni-salt ANSH as the most prominent example), oxides (REE oxides), or ecocatalysts (Eco-Mn, Eco-Ni, Eco-Zn), as well as hydro- and biochars (As, Zn), were successfully generated from biomass of different hyperaccumulator species, showing potential for application in different industries and increasing economic feasibility of agromining chain. There is still ongoing search for cost-effective recovery methods, with small environmental footprint, use of safe chemicals, and within circular economy approach. Generally, economic feasibility of agromining depends on the existence of the suitable area for agromining development, use of hyperaccumulator species with increased biomass yield, accumulation level of selected metal, use of adequate agrotechnologies for simultaneous promotion of biomass yield and metal uptake enhancement, cost of metal recovery technology, and current value of metal at the market.

While some metallic elements or their compounds reach high prices on worlds market, others are cheaper and more readily available from existing primary sources. In that sense, recovery of Ni, REEs or PGMs shows increased potential to become alternative to traditional mining in sites where they are present in low grade, while other elements (such as As, Cr, or Tl) are more suited primarily for reducing of pollution level and increasing the soil quality, whereas element recovery is of secondary importance. There is number of phytomining studies conducted in laboratory and greenhouse conditions; however, there is considerable lack of field studies that would help adjust agromining practices and technologies to the real-world applications. Similarly, even though metal recovery from accumulating plants was tested on laboratory scale, upscaling to industrial level often requires design of special equipment or facility, taking into consideration biomass type, metal content, and technology applied. Overall, although additional research are needed to address present limitations, agromining has a potential to be developed into alternative mining technology for obtaining strategic and critical raw materials from plant resources, thus contributing to the diversification of supply chains and greater stability needed in sustainable development of clean technologies.

References

Abbas, G., Murtaza, B., Bibi, I., Shahid, M., Niazi, N.K., Khan, M.I., et al., 2018. Arsenic uptake, toxicity, detoxification, and speciation in plants: physiological, biochemical, and molecular aspects. International Journal of Environmental Research and Public Health 15 (1), 59. Available at https://doi.org/10.3390/ijerph15010059, http://www.mdpi.com/1660-4601/15/1/59/pdf.

Anderson, C.W.N., Brooks, R.R., Stewart, R.B., Simcock, R., 1998. Harvesting a crop of gold in plants. Nature 395 (6702), 553–554. Available at https://doi.org/10.1038/26875.

Anderson, C.W.N., Brooks, R.R., Chiarucci, A., LaCoste, C.J., Leblanc, M., Robinson, B.H., et al., 1999. Phytomining for nickel, thallium and gold. Journal of Geochemical Exploration 67 (1–3), 407–415. Available at https://doi.org/10.1016/s0375-6742(99)00055-2.

Anderson, C., Moreno, F., Meech, J., 2005. A field demonstration of gold phytoextraction technology. Minerals Engineering 18 (4), 385–392. Available at https://doi.org/10.1016/j.mineng.2004.07.002.

Azizi, M., Faz, A., Zornoza, R., Martinez-Martinez, S., Acosta, J.A., 2023. Phytoremediation potential of native plant species in mine soils polluted by metal (loid)s and rare earth elements.Plants 12 (6), 1219. Available at https://www.mdpi.com/journal/plants.

Baker, A.J.M., Brooks, R.R., 1989. Terrestrial higher plants which hyperaccumulate metallic elements: a review of their distribution, ecology and phytochemistry. Biorecovery 1, 81–126.

Bani, A., Echevarria, G., Sulçe, S., Morel, J.L., 2015. Improving the agronomy of *Alyssum murale* for extensive phytomining: a five-year field study. International Journal of Phytoremediation 17 (2), 117–127. Available at https://doi.org/10.1080/15226514.2013.862204, http://www.tandf.co.uk/journals/titles/15226514.asp.

Barbaroux, R., Plasari, E., Mercier, G., Simonnot, M.-O., Morel, J.-L., Blais, J.-F., 2012. A new process for nickel ammonium disulfate productionfrom ash of the hyperaccumulating plant Alyssum murale. Science of The Total Environment 423, 111–119. Available at https://doi.org/10.1016/j.scitotenv.2012.01.063.

Battke, F., Leopold, K., Maier, M., Schmidhalter, U., Schuster, M., 2008. Palladium exposure of barley: uptake and effects. Plant Biology 10 (2), 272–276. Available at https://doi.org/10.1111/j.1438-8677.2007.00017.x.

Bert, V., Bonnin, I., Saumitou-Laprade, P., De Laguérie, P., Petit, D., 2002. Do *Arabidopsis halleri* from nonmetallicolous populations accumulate zinc and cadmium more effectively than those from metallicolous populations? New Phytologist 155 (1), 47–57. Available at https://doi.org/10.1046/j.1469-8137.2002.00432.x.

Bihanic, C., Richards, K., Olszewski, T.K., Grison, C., 2020. Eco-Mn ecocatalysts: toolbox for sustainable and green Lewis acid catalysis and oxidation reactions. ChemCatChem 12 (6), 1529–1545. Available at https://doi.org/10.1002/cctc.201901845, http://onlinelibrary.wiley.com/journal/10.1002/(ISSN)1867-3899.

Boominathan, R., Saha-Chaudhury, N.M., Sahajwalla, V., Doran, P.M., 2004. Production of nickel bio-ore from hyperaccumulator plant biomass: applications in phytomining. Biotechnology and Bioengineering 86 (3), 243–250. Available at https://doi.org/10.1002/bit.10795.

Borovička, J., Řanda, Z., Jelínek, E., Kotrba, P., Dunn, C.E., 2007. Hyperaccumulation of silver by *Amanita strobiliformis* and related species of the section Lepidella. Mycological Research 111 (11), 1339–1344. Available at https://doi.org/10.1016/j.mycres.2007.08.015.

Brooks, R.R., Chambers, M.F., Nicks, L.J., Robinson, B.H., 1998. Phytomining. Trends in Plant Science 3 (9), 359–362. Available at https://doi.org/10.1016/s1360-1385(98)01283-7.

Chai, Y., Chen, A., Bai, M., Peng, L., Shao, J., Yuan, J., et al., 2022. Valorization of heavy metal contaminated biomass: recycling and expanding to functional materials. Journal of Cleaner Production 366, 132771. Available at https://doi.org/10.1016/j.jclepro.2022.132771.

Chaney, R., 1983. Plant uptake of inorganic waste. In: Parr, J.E. (Ed.), Land Treatment of Hazardous Waste. Noyes Data Corp., Park Ridge, pp. 50–76.

Chaney, R.L., Baker, A.J.M., Morel, J.L., 2018. The long road to developing agromining/phytomining. In: van der Ent, A., Echevarria, G., Baker, A.J.M., Morel, J.L. (Eds.), Agromining: Farming for Metals: Extracting Unconventional Resources Using Plants. Springer International Publishing, Cham, pp. 1–17.

Chaney, R.L., Angle, J.S., Baker, A.J.M., Li, M., 1998. Method for phytomining of Ni, cobalt and other metals from soil. U.S. Patent 7845711.

Chen, P., Ilton, E.S., Wang, Z., Rosso, K.M., Zhang, X., 2024. Global rare earth element resources: a concise review. Applied Geochemistry 175, 106158. Available at https://doi.org/10.1016/j.apgeochem.2024.106158.

Chen, T., Lei, M., Wan, X., Zhou, X., Yang, J., Guo, G., et al., 2018. Element case studies: arsenic. In: van der Ent, A., Baker, A.J., Echevarria, G., Simonnot, M.O., Morel, J.L. (Eds.), Agromining: Farming for Metals. Mineral Resource Reviews. Springer, Cham, pp. 275–281. Available at http://doi.org/10.1007/978-3-319-61899-9_17.

Chen, W., Dai, Y., Liu, Z., Zhang, H., 2024. The evolution of global zinc trade network: patterns and implications. Resources Policy 90, 104727. Available at https://doi.org/10.1016/j.resourpol.2024.104727.

Chour, Z., Laubie, B., Morel, J.L., Tang, Y., Qiu, R., Simonnot, M.O., et al., 2018. Recovery of rare earth elements from *Dicranopteris dichotoma* by an enhanced ion exchange leaching process. Chemical Engineering and Processing - Process Intensification 130, 208–213. Available at https://doi.org/10.1016/j.cep.2018.06.007, https://www.elsevier.com/inca/publications/store/5/0/4/0/8/1/index.htt.

Chour, Z., Laubie, B., Morel, J.L., Tang, Y.T., Simonnot, M.O., Muhr, L., 2020. Basis for a new process for producing REE oxides from *Dicranopteris linearis*. Journal of Environmental Chemical Engineering 8 (4), 103961. Available at https://doi.org/10.1016/j.jece.2020.103961, http://www.journals.elsevier.com/journal-of-environmental-chemical-engineering/.

Corzo Remigio, A., Pošćić, F., Nkrumah, P.N., Edraki, M., Spiers, K.M., Brueckner, D., et al., 2022. Comprehensive insights in thallium ecophysiology in the hyperaccumulator *Biscutella laevigata*. Science of the Total Environment 838 (Part 2), 155899. Available at https://doi.org/10.1016/j.scitotenv.2022.155899, https://www.elsevier.com/locate/scitotenv.

Cui, X., Shen, Y., Qianying, Y., Kawi, S., He, Z., Yang, X., et al., 2018. Simultaneous syngas and biochar production during heavy metal separation from Cd/Zn hyperaccumulator (*Sedum alfredii*) by gasification. Chemical Engineering Journal 347, 543–551. Available at https://doi.org/10.1016/j.cej.2018.04.133.

da Silva, E.B., de Oliveira, L.M., Wilkie, A.C., Liu, Y., Ma, L.Q., 2018. Arsenic removal from As-hyperaccumulator *Pteris vittata* biomass: coupling extraction with precipitation. Chemosphere 193, 288–294. Available at https://doi.org/10.1016/j.chemosphere.2017.10.116, https://www.elsevier.com/locate/chemosphere.

Dilshara, P., Abeysinghe, B., Premasiri, R., Dushyantha, N., Ratnayake, N., Senarath, S., et al., 2024. The role of nickel (Ni) as a critical metal in clean energy transition: applications, global distribution and occurrences, production-demand and phytomining. Journal of Asian Earth Sciences 259, 105912. Available at https://doi.org/10.1016/j.jseaes.2023.105912, http://www.sciencedirect.com/science/journal/13679120.

Dinh, T., Kovács, H., Dobó, Z., 2024. The formation of gold in woody biomass combustion ashes. Heliyon 10 (11), e32425. Available at https://doi.org/10.1016/j.heliyon.2024.e32425.

Dresler, S., Hawrylak-Nowak, B., Strzemski, M., Wójciak-Kosior, M., Sowa, I., Hanaka, A., et al., 2019. Metabolic changes induced by silver ions in *Carlina acaulis*. Plants 8 (11), 517. Available at https://doi.org/10.3390/plants8110517.

Elmayel, I., Esbrí, J.M., Efrén, G.O., García-Noguero, E.M., Elouear, Z., Jalel, B., et al., 2020. Evolution of the speciation and mobility of Pb, Zn and Cd in relation to transport processes in a mining environment. International journal of environmental research and public health 17 (14), 1–16. Available at https://doi.org/10.3390/ijerph17144912, https://www.mdpi.com/1660-4601/17/14/4912/pdf.

Escande, V., Olszewski, T.K., Grison, C., 2014. Preparation of ecological catalysts derived from Zn hyperaccumulating plants and their catalytic activity in Diels–Alder reaction. Comptes Rendus. Chimie. 17 (7–8), 731–737. Available at https://doi.org/10.1016/j.crci.2013.09.009.

Escande, V., Olszewski, T.K., Grison, C., 2015. From biodiversity to catalytic diversity: how to control the reaction mechanism by the nature of metallophytes. Environmental Science and Pollution Research 22 (8), 5653–5666. Available at https://doi.org/10.1007/s11356-014-3483-6, http://www.springerlink.com/content/0944-1344.

Escande, V., Poullain, C., Clavé, G., Petit, E., Masquelez, N., Hesemann, P., et al., 2017. Bio-based and environmental input for transfer hydrogenation using EcoNi(0) catalyst in isopropanol. Applied Catalysis B: Environmental 210, 495–503. Available at https://doi.org/10.1016/j.apcatb.2017.04.023.

European Commission. 2023. Study on the Critical Raw Materials for the EU 2023 – Final Report.

Eze, V.C., Harvey, A.P., 2018. Extractive recovery and valorisation of arsenic from contaminated soil through phytoremediation using Pteris cretica. Chemosphere 208, 484–492. Available at https://doi.org/10.1016/j.chemosphere.2018.06.027, www.elsevier.com/locate/chemosphere.

Feichtmeier, N., Leopold, K., 2014. Bioavailability of platinum group elements to plants—a review. In: Zereini, F., Wiseman, C. (Eds.), Platinum Metals in the Environment. Environmental Science and Engineering. Springer, Berlin, Heidelberg. Available at https://doi.org/10.1007/978-3-662-44559-4_20.

Fernando, D.R., Bakkaus, E.J., Perrier, N., Baker, A.J.M., Woodrow, I.E., Batianoff, G.N., et al., 2006. Manganese accumulation in the leaf mesophyll of four tree species: a PIXE/EDAX localization study. New Phytologist 171 (4), 751–758. Available at https://doi.org/10.1111/j.1469-8137.2006.01783.x.

Genchi, G., Carocci, A., Lauria, G., Sinicropi, M.S., Catalano, A., 2021. Thallium use, toxicity, and detoxification therapy: an overview. Applied Sciences 11 (18), 8322. Available at https://doi.org/10.3390/app11188322.

Gilkes, R.J., McKenzie, R.M., 1988. Geochemistry and mineralogy of manganese in soils. In: Graham, R.D., Hannam, R.J., Uren, N.C. (Eds.), Manganese in Soils and Plants. vol 33: Developments in Plant and Soil Sciences. Springer, Dordrecht. Available at https://doi.org/10.1007/978-94-009-2817-6_3.

Grison, C., Ki, Y., 2021. Ecocatalysis: a new vision of green and sustainable chemistry. Current Opinion in Green and Sustainable Chemistry 100461. Available at https://doi.org/10.1016/j.cogsc.2021.100461.

Grison, C., Velati, A., Escande, V., Grison, C., 2015. Metallophytes for organic synthesis: towards new bio-based selective protection/deprotection procedures. Environmental

Science and Pollution Research; Combining Phytoextraction and Ecological Catalysis: an Environmental, Ecological, Ethic and Economic Opportunity 22 (8), 5686–5698. Available at 10.1007/s11356-014-3526-zf.

Grison, C., Escande, V., Petit, E., Garoux, L., Boulanger, C., Grison, C., 2013. *Psychotria douarrei* and *Geissois pruinosa*, novel resources for the plant-based catalytic chemistry. RSC Advances 3 (44), 22340. Available at https://doi.org/10.1039/C3RA43995J.

Grison, C.M., Mazel, M., Sellini, A., Escande, V., Biton, J., Grison, C., 2015. The leguminous species Anthyllis vulneraria as a Zn-hyperaccumulator and eco-Zn catalyst resources. Environmental Science and Pollution Research 22 (8), 5667–5676. Available at https://doi.org/10.1007/s11356-014-3605-1.

Grison, C., Escarre, J., 2012. Use of metal-accumulating plants for the preparation of catalysts that can be used in chemical reactions. US Patent US20120316340A1.

Grosjean, N., Le Jean, M., Berthelot, C., Chalot, M., Gross, E.M., Blaudez, D., 2019. Accumulation and fractionation of rare earth elements are conserved traits in the *Phytolacca* genus. Scientific Reports 9, 18458. Available at https://doi.org/10.1038/s41598-019-54238-3.

Guilpain, M., Laubie, B., Zhang, X., Morel, J.L., Simonnot, M.O., 2018. Speciation of nickel extracted from hyperaccumulator plants by water leaching. Hydrometallurgy 180, 192–200. Available at https://doi.org/10.1016/j.hydromet.2018.07.024.

Harris, A.T., Bali, R., 2007. On the formation and extent of uptake of silver nanoparticles by live plants. Journal of Nanoparticle Research 10 (4), 691–695. Available at https://doi.org/10.1007/s11051-007-9288-5.

Harris, B., White, C., 2019. Method for the recovery of manganese products from various feedstocks. Patent WO2019161447A1.

Harumain, Z.A.S., Parker, H.L., Muñoz García, A., Austin, M.J., McElroy, C.R., Hunt, A.J., et al., 2017. Toward financially viable phytoextraction and production of plant-based palladium catalysts. Environmental Science and Technology 51 (5), 2992–3000. Available at https://doi.org/10.1021/acs.est.6b04821, http://pubs.acs.org/journal/esthag.

Hazotte, C., Laubie, B., Rees, F., Morel, J.L., Simonnot, M.O., 2017. A novel process to recover cadmium and zinc from the hyperaccumulator plant Noccaea caerulescens. Hydrometallurgy 174, 56–65. Available at https://doi.org/10.1016/j.hydromet.2017.09.012.

Hazotte, C., Laubie, B., Pacault, S., Dufaud, O., Simonnot, M.- O., 2020. Evaluation of the performance of nickel hyperaccumulator plants as combustion fuel. Biomass and Bioenergy 140, 105671. Available at https://doi.org/10.1016/j.biombioe.2020.105671.

Houzelot, V., Ranc, B., Laubie, B., Simonnot, M.-O., 2018. Agromining of hyperaccumulator biomass: study of leaching kinetics ofextraction of nickel, magnesium, potassium, phosphorus, iron, and manganese from *Alyssum murale* ashes by sulfuric acid. Chemical Engineering Research and Design 129, 1–11. Available at https://doi.org/10.1016/j.cherd.2017.10.030.

International Nickel Study Group. 2025. Available at https://insg.org/index.php/about-nickel/production-usage/

Jaffré, T., 1979. Accumulation du manganèse par les Protéacées de Nouvelle - Calédonie. Comptes Rendus de l'Académie des Sciences. Série D: Sciences Naturelles 289, 425–428.

Jaffré, T., Brooks, R.R., Lee, J., Reeves, R.D., 1976. *Sebertia acuminata*: a hyperaccumulator of nickel from New Caledonia. Science 193 (4253), 579–580. Available at https://doi.org/10.1126/science.193.4253.579.

Jakovljević, K., Mišljenović, T., Savović, J., Ranković, D., Ranđelović, D., Mihailović, N., et al., 2020. Accumulation of trace elements in Tussilago farfara colonizing post-flotation tailing sites in Serbia. Springer, Serbia Environmental Science and Pollution Research 27 (4), 4089–4103. Available at https://doi.org/10.1007/s11356-019-07010-z, https://link.springer.com/journal/11356.

Jakovljević, K., Bani, A., Pavlova, D., Konstantinou, M., Dimitrakopoulos, P.G., Kyrkas, D., et al., 2022. Hyperaccumulator plant discoveries in the Balkans: accumulation, distribution, and practical applications. Institute of Botany and Botanical Garden "Jevremovac", University of Belgrade, Serbia Botanica Serbica 46 (2), 161–178. Available at https://doi.org/10.2298/BOTSERB2202161J, http://botanicaserbica.bio.bg.ac.rs/.

Jakovljević, K., Mišljenović, T., Bačeva Andonovska, K., Echevarria, G., Baker, A.J.M., Brueckner, D., et al., 2023. Thallium hyperaccumulation status of the violets of the Allchar arsenic-thallium deposit (North Macedonia) confirmed through synchrotron μXRF imaging. Metallomics 15 (11). Available at https://doi.org/10.1093/mtomcs/mfad063, https://academic.oup.com/metallomics/issue/13/1.

Jally, B., Laubie, B., Tang, Y.T., Simonnot, M.O., 2021. Processing of plants to products: gold, REEs and other elements. In: van der Ent, A., Baker, A.J., Echevarria, G., Simonnot, M.O., Morel, J.L. (Eds.), Agromining: Farming for Metals. Mineral Resource Reviews. Springer, Cham, pp. 63–74. Available at https://doi.org/10.1007/978-3-030-58904-2_4.

Jally, B., Laubie, B., Chour, Z., Muhr, L., Qiu, R., Morel, J.L., et al., 2021. A new method for recovering rare earth elements from the hyperaccumulating fern *Dicranopteris linearis* from China. Minerals Engineering 166, 106879. Available at https://doi.org/10.1016/j.mineng.2021.106879, http://www.journals.elsevier.com/minerals-engineering/.

Jiang, H., Chen, X., Chen, S., Li, H., Peng, Y., Zhu, A., et al., 2022. Recovery of arsenic and practical utilization of aqueous phase in hydrothermal liquefaction of hyperaccumulator. Chemical Engineering Journal 439, 135514. Available at https://doi.org/10.1016/j.cej.2022.135514.

Jovanović, G., Ranđelović, D., Marković, B., Sokić, M., 2022. Overview of technologies for Zn extraction from hyperaccumulating plants: current state of research and future directions. Journal of Mining and Metallurgy A: Mining 58 (1), 29–38. Available at https://doi.org/10.5937/jmma2201029j.

Kabata-Pendias, A., 2011. Trace Elements in Soils and Plants, 4th ed. Taylor & Francis, New York, pp. 548.

Kania, H., Saternus, M., 2023. Evaluation and current state of primary and secondary zinc production—a review. Applied Sciences 13 (3), 2003. Available at https://doi.org/10.3390/app13032003.

Karczewska, A., Bogda, A., Krysiak, A., 2007. Arsenic in soils in the areas of former mining and mineral processing in Lower Silesia, southwestern Poland. Arsenic in Soil and Groundwater Environment - Biogeochemical Interactions, Health Effects and Remediation 411–440. Available at https://doi.org/10.1016/s1875-1121(06)09016-x.

Kaya, M., Hussaini, S., Kursunoglu, S., 2020. Critical review on secondary zinc resources and their recycling technologies. Hydrometallurgy 195, 105362. Available at https://doi.org/10.1016/j.hydromet.2020.105362.

Kenney, K.L., Smith, W.A., Gresham, G.L., Westover, T.L., 2013. Understanding biomass feedstock variability. Biofuels 4 (1), 111–127. Available at https://doi.org/10.4155/bfs.12.83.

Kikis, C., Thalassinos, G., Antoniadis, V., 2024. Soil phytomining: recent developments—a review. Soil Systems 8 (1), 8. Available at https://doi.org/10.3390/soilsystems8010008.

Kolesnikov, S.I., Tsepina, N.I., Sudina, L.V., Minnikova, T.V., Kazeev, K.S., Akimenko, Y.V., 2020. Silver ecotoxicity estimation by the soil state biological indicators. Applied and Environmental Soil Science 2020, 1–9. Available at https://doi.org/10.1155/2020/1207210.

LaCoste, C., Robinson, B., Brooks, R., Anderson, C., Chiarucci, A., Leblanc, M., 1999. The phytoremediation potential of thallium-contaminated soils using *Iberis* and *Biscutella* species. International Journal of Phytoremediation 1 (4), 327–338. Available at https://doi.org/10.1080/15226519908500023.

Lamb, A., Anderson, C., Haverkamp, R., 2001b. The induced accumulation of gold in the plants Brassica juncea, Berkheya coddii and chicory. Chemistry in New Zealand 65 (2), 34–36.

Lamb, A., Anderson, C., Haverkamp, R., 2001. The extraction of gold from plants and its application to phytomining. Chemistry in New Zealand 65, 31–33.

Leblanc, M., Petit, D., Deram, A., Robinson, B.H., Brooks, R.R., 1999. The phytomining and environmental significance of hyperaccumulation of thallium by *Iberis intermedia* from southern France. Economic Geology 94 (1), 109–113. Available at https://doi.org/10.2113/gsecongeo.94.1.109.

Lei, M., Dong, Z., Jiang, Y., Longhurst, P., Wan, X., Zhou, G., 2019. Reaction mechanism of arsenic capture by a calcium-based sorbent during the combustion of arsenic-contaminated biomass: a pilot-scale experience. Frontiers of Environmental Science & Engineering 13 (2), 24. Available at https://doi.org/10.1007/s11783-019-1110-y.

Li, J.T., Gurajala, H.K., Wu, L.H., Van Der Ent, A., Qiu, R.L., Baker, A.J.M., et al., 2018. Hyperaccumulator plants from China: a synthesis of the current state of knowledge. Environmental Science and Technology 52 (21), 11980–11994. Available at https://doi.org/10.1021/acs.est.8b01060, http://pubs.acs.org/journal/esthag.

Li, S., Zhang, W., 2024. REE recovery and hydrochar production from a hyperaccumulator by microwave-assisted hydrothermal carbonization. Minerals Engineering 208, 108595. Available at https://doi.org/10.1016/j.mineng.2024.108595.

Li, S., Ji, B., Zhang, W., 2023. A review on the thermochemical treatments of biomass: implications for hydrochar production and rare earth element recovery from hyperaccumulators. Chemosphere 342, 140140. Available at 10.1016/j.chemosphere.2023.140140.

Liu, C., Liu, W.S., van der Ent, A., Morel, J.L., Zheng, H.X., Wang, G.B., et al., 2021. Simultaneous hyperaccumulation of rare earth elements, manganese and aluminum in *Phytolacca americana* in response to soil properties. Chemosphere 282, 131096. Available at https://doi.org/10.1016/j.chemosphere.2021.131096, https://www.elsevier.com/locate/chemosphere.

Losfeld, G., de La Blache, P.V., Escande, V., Grison, C., 2012b. Zinc hyperaccumulating plants as renewable resources for the chlorination process of alcohols. Green Chemistry Letters and Reviews 5 (3), 451–456. Available at https://doi.org/10.1080/17518253.2012.667157.

Losfeld, G., de la Blache, P.V., Escande, V., Grison, C., 2012. Lewis acid catalysts: a potential exploitation for zinc and nickel phytoextraction. WIT Transactions on Ecology and the Environment 162, 563–575. Available at https://doi.org/10.2495/EID120491, http://www.library.witpress.com/pages/listBooks.asp?tID=4.

Ma, L.Q., Komar, K.M., Tu, C., Zhang, W.H., Cai, Y., Kennelley, E.D., 2001. A fern that hyperaccumulates arsenic: a hardy, versatile, fast-growing plant helps to remove arsenic from contaminated soils. Nature 409, 579. Available at https://doi.org/10.1038/35054664 –579.

Marković, B., Ranđelović, D., Jovanović, G., Tomović, G., Jakovljević, K., Mišljenović, T., et al., 2021. Extraction of ammonium nickel sulfate hexahydrate by hydrometallurgical process from the hyperaccumulating plant odontarrhena muralis – Case study from Serbia. Hemijska Industrija 75 (5), 285–296. Available at https://doi.org/10.2298/HEMIND210701027M, http://www.doiserbia.nb.rs/Article.aspx?ID=0367-598X2105285M.

Marković, B., Jovanović, G., Ranđelović, D., Miletić, M., Vuković, N., Vujović, N., et al., 2025. Impact of recrystallization on the purity of ANSH crystals produced from Ni hyperaccumulator plant. Tehnika 80 (1), 45–53. Available at https://doi.org/10.5937/tehnika2501045m.

Masinire, F., Adenuga, D.O., Tichapondwa, S.M., Chirwa, E.M.N., 2022. Palladium phytoremediation and phytomining potential of vetiver grass (*Chrysopogon zizanioides*).

Italian Association of Chemical Engineering - AIDIC, South Africa. Chemical Engineering Transactions 94, 1429–1434. Available at https://www.aidic.it/cet/22/94/238.pdf.

Meharg, A.A., 2003. Variation in arsenic accumulation: hyperaccumulation in ferns and their allies. New Phytologist 157 (1), 25–31. Available at https://doi.org/10.1046/j.1469-8137.2003.00541.x.

Mercier, G., Barbaroux, R., Plasari, E., Blais, J.-F., Simonnot, M.-O., Morel, J.-L., 2012. Procédé de production d'un sel de sulfate double de nickel et d'ammonium à partir de plantes hyperaccumulatrices. Patent WO2012/103651.

Millaleo, R., Reyes- Diaz, M., Ivanov, A., Mora, M., Alberdi, M., 2010. Manganese as essential and toxic element for plants: transport, accumulation and resistance mechanisms. Journal of Soil Science and Plant Nutrition 10 (4), 470–481. Available at https://doi.org/10.4067/s0718-95162010000200008.

Milner, M.J., Kochian, L.V., 2008. Investigating heavy-metal hyperaccumulation using *Thlaspi caerulescens* as a model system. Annals of Botany 102 (1), 3–13. Available at https://doi.org/10.1093/aob/mcn063.

Nemutandani, T., Dutertre, D., Chimuka, L., Cukrowska, E., Tutu, H., 2006. The potential of *Berkheya coddii* for phytoextraction of nickel, platinum, and palladium contaminated sites. Toxicological & Environmental Chemistry 88 (2), 175–185. Available at https://doi.org/10.1080/02772240600585842.

Nkrumah, P.N., Baker, A.J.M., Chaney, R.L., Erskine, P.D., Echevarria, G., Morel, J.L., et al., 2016. Current status and challenges in developing nickel phytomining: an agronomic perspective. Springer International Publishing, Australia Plant and Soil 406 (1-2), 55–69. Available at https://doi.org/10.1007/s11104-016-2859-4, http://www.wkap.nl/journalhome.htm/0032-079X.

Nkrumah, P.N., Tisserand, R., Chaney, R.L., Baker, A.J.M., Morel, J.L., Goudon, R., et al., 2019. The first tropical 'metal farm': some perspectives from field and pot experiments. Journal of Geochemical Exploration 198, 114–122. Available at https://doi.org/10.1016/j.gexplo.2018.12.003.

Odjegba, V.J., Brown, M.T., Turner, A., 2007. Studies on the effects of platinum group elements on *Lactuca sativa* L. American Journal of Plant Physiology 2 (3), 183–194. Available at https://doi.org/10.3923/ajpp.2007.183.194.

Oleszek, S., Shiota, K., Chen, M., Takaoka, M., 2022. Effective Separation and Recovery of Manganese and Potassium from Biomass Ash by Solvent Extraction. ACS Omega 7 (23), 20155–20164. Available at https://doi.org/10.1021/acsomega.2c02002.

Pandey, V.C., Ancona, V., Roy, M., Ranđelović, D., 2023. Aromatic plant-based phytoremediation, socio-economic and agricultural. Sustainability, 1st ed. Elsevier ISBN 9780443190827.

Parker, H.L., Rylott, E.L., Hunt, A.J., Dodson, J.R., Taylor, A.F., Bruce, N.C., et al., 2014. Supported palladium nanoparticles synthesized by living plants as a catalyst for Suzuki-Miyaura reactions. PLoS ONE 9 (1). Available at https://doi.org/10.1371/journal.pone.0087192, http://www.plosone.org/article/fetchObject.action?uri=info%3Adoi%2F10.1371%2Fjournal.pone.0087192&representation=PDF.

Peterson, P.J., Girling, C.A., 1981. Other trace metals: chromium. In: In: Lepp, N.W. (Ed.), Effect of Heavy Metal on Pollution on Plants, vol. I. Applied Science Publishers, London, pp. 222–229.

Pilchin, A., Eppelbaum, L., 2017. Concentration of platinum group elements during the early Earth evolution: a review. Natural Resources 08 (03), 172–233. Available at https://doi.org/10.4236/nr.2017.83012.

Pitcairn, I.K., 2013. Background concentrations of gold in different rock types. Applied Earth Science 120 (1), 31–38. Available at https://doi.org/10.1179/1743275811y.0000000021.

Punshon, T., Jackson, B.P., Meharg, A.A., Warczack, T., Scheckel, K., Guerinot, M.L., 2017. Understanding arsenic dynamics in agronomic systems to predict and prevent uptake by crop plants. Science of the Total Environment 581-582, 209–220. Available at https://doi.org/10.1016/j.scitotenv.2016.12.111, https://www.elsevier.com/locate/scitotenv.

Qian, F., Zhu, X., Liu, Y., Shi, Q., Wu, L., Zhang, S., et al., 2018. Influences of temperature and metal on subcritical hydrothermal liquefaction of hyperaccumulator: implications for the recycling of hazardous hyperaccumulators. Environmental Science & Technology 52 (4), 2225–2234. Available at https://doi.org/10.1021/acs.est.7b03756.

Qu, J., Luo, C., Cong, Q., Yuan, X., 2012. Carbon nanotubes and Cu–Zn nanoparticles synthesis using hyperaccumulator plants. Environmental Chemistry Letters 10 (2), 153–158. Available at https://doi.org/10.1007/s10311-011-0335-1.

Ranđelović, D., Pandey, V.C., 2023. Bioenergy crop-based ecological restoration of degraded land. In: Pandey, V.C. (Ed.), Bio-Inspired Land Remediation. Springer, pp. 1–29. (Chapter 1). Available at 10.1007/978-3-031-04931-6_1.

Ranđelović, D., Mutić, J., Marjanović, P., Đorđević, T., Kašanin-Grubin, M., 2020. Geochemical distribution of selected elements in flotation tailings and soils/sediments from the dam spill at the abandoned antimony mine Stolice, Serbia. Environmental Science and Pollution Research 27 (6), 6253–6268. Available at https://doi.org/10.1007/s11356-019-07348-4.

Ravindra, K., Bencs, L., Van Grieken, R., 2004. Platinum group elements in the environment and their health risk. Science of the Total Environment 318 (1–3), 1–43. Available at https://doi.org/10.1016/s0048-9697(03)00372-3.

Reeves, R.D., Schwartz, C., Morel, J.L., Edmondson, J., 2001. Distribution and metal-accumulating behavior of Thlaspi caerulescens and associated metallophytes in France. International journal of phytoremediation 3 (2), 145–172. Available at https://doi.org/10.1080/15226510108500054, http://www.tandf.co.uk/journals/titles/15226514.asp.

Reeves, R.D., Baker, A.J.M., Jaffré, T., Erskine, P.D., Echevarria, G., van der Ent, A., 2018. A global database for plants that hyperaccumulate metal and metalloid trace elements. New Phytologist 218 (2), 407–411. Available at https://doi.org/10.1111/nph.14907, http://onlinelibrary.wiley.com/journal/10.1111/(ISSN)1469-8137.

Rehman, M.U., Khan, R., Khan, A., Qamar, W., Arafah, A., Ahmad, A., et al., 2021. Fate of arsenic in living systems: implications for sustainable and safe food chains. Journal of Hazardous Materials 417, 126050. Available at https://doi.org/10.1016/j.jhazmat.2021.126050.

Robinson, B., Anderson, C., 2018. Agromining of thallium and noble metals. In: van der Ent, A., Echevarria, G., Baker, A.J.M., Morel, J.L. (Eds.), Agromining: Farming for Metals: Extracting Unconventional Resources Using Plants. Springer International Publishing, Cham, pp. 253–261. Available at https://doi.org/10.1007/978-3-319-61899-9_15.

Robinson, B.H., Chiarucci, A., Brooks, R.R., Petit, D., Kirkman, J.H., Gregg, P.E.H., et al., 1997. The nickel hyperaccumulator plant *Alyssum bertolonii* as a potential agent for phytoremediation and phytomining of nickel. Journal of Geochemical Exploration 59 (2), 75–86. Available at https://doi.org/10.1016/s0375-6742(97)00010-1.

Robinson, B., Anderson, C., 2021. Agromining of Thallium and Noble Metals. In: van der Ent, A., Baker, A.J., Echevarria, G., Simonnot, M.-O., Morel, J.L. (Eds.), Agromining: Farming for Metals. 2021, Mineral Resource Reviews. Springer, Cham. Available at https://doi.org/10.1007/978-3-030-58904-2_20.

Rodrigues, J., Houzelot, V., Ferrari, F., Echevarria, G., Laubie, B., Morel, J.L., et al., 2016. Life cycle assessment of agromining chain highlights role of erosion control and bioenergy. Journal of Cleaner Production 139, 770–778. Available at https://doi.org/10.1016/j.jclepro.2016.08.110.

Rodriguez, E., Parsons, J.G., Peralta-Videa, J.R., Cruz-Jimenez, G., Romero-Gonzalez, J., Sanchez-Salcido, B.E., et al., 2007. Potential of *Chilopsis linearis* for gold phytomining: using XAS to determine gold reduction and nanoparticle formation within plant tissues. International Journal of Phytoremediation 9 (2), 133–147. Available at https://doi.org/10.1080/15226510701232807.

Rudnick, R.L., Gao, S., 2003. The composition of the continental crust. In: Holland, H.D., Turekian, K.K. (Eds.), Treatise on Geochemistry. vol. 3: The Crust. Elsevier-Pergamon, Oxford, pp. 1–64. Available at http://doi.org/10.1016/b0-08-043751-6/03016-4.

Rylott, E.L., van der Ent, A., 2025. Harnessing hyperaccumulator plants to recover technology-critical metals: where are we at? New Phytologist 246 (3), 859–866. Available at https://doi.org/10.1111/nph.20449.

Sagiroglu, A., Sasmaz, A., Sen, O., 2006. Hyperaccumulator plants of the Keban mining district and their possible impact on the environment. Polish Journal of Environmental Studies 15 (2), 317–325.

Salinitro, M., Isnard, S., Brueckner, D., Spiers, K.M., Aarts, M.G.M., Remigio, A.C., et al., 2024. Uncovering the physiology and distribution of thallium in Tl-hyperaccumulating and Tl-sensitive populations of Biscutella laevigata L. Annals of Botany 134 (5), 803–814. Available at https://doi.org/10.1093/aob/mcae115, https://www.aob.oxfordjournals.org/.

Salminen, R., Batista, M.J., Bidovec, M., Demetriades, A., Vivo, B.De, et al., 2005. Geochemical atlas of Europe. Part 1: Background information, methodology and maps. Geological Survey of Finland. Available at http://weppi.gtk.fi/publ/foregsatlas/articles.php.

Saper, R.B., Rash, R., 2009. Zinc: an essential micronutrient. American Family Physician 79 (9), 768–772. Available at https://www.aafp.org/afp/20090501/768.pdf.

Shahid, M., Shamshad, S., Rafiq, M., Khalid, S., Bibi, I., Niazi, N.K., et al., 2017. Chromium speciation, bioavailability, uptake, toxicity and detoxification in soil-plant system: a review. Chemosphere 178, 513–533. Available at https://doi.org/10.1016/j.chemosphere.2017.03.074, https://www.elsevier.com/locate/chemosphere.

Shan, X., Wang, H., Zhang, S., Zhou, H., Zheng, Y., Yu, H., et al., 2003. Accumulation and uptake of light rare earth elements in a hyperaccumulator *Dicropteris dichotoma*. Plant Science 165, 1343–1353. Available at https://doi.org/10.1016/S0168-9452(03)00361-3.

Sharma, N.C., Sahi, S.V., Nath, S., Parsons, J.G., Gardea-Torresdey, J.L., Tarasankar, P., 2007. Synthesis of plant-mediated gold nanoparticles and catalytic role of biomatrix-embedded nanomaterials. Environmental Science and Technology 41 (14), 5137–5142. Available at https://doi.org/10.1021/es062929a.

Sharma, P., Singh, S.P., Tripathi, R.D., Tong, Y.W., 2023. Chromium toxicity and tolerance mechanisms in plants through cross-talk of secondary messengers: an overview of pathways and mechanisms. Environmental Pollution 320, 121049. Available at https://doi.org/10.1016/j.envpol.2023.121049, https://www.journals.elsevier.com/environmental-pollution.

Stein, R.J., Höreth, S., de Melo, J.R.F., Syllwasschy, L., Lee, G., Garbin, M.L., et al., 2017. Relationships between soil and leaf mineral composition are element-specific, environment-dependent and geographically structured in the emerging model Arabidopsis halleri. New Phytologist 213 (3), 1274–1286. Available at https://doi.org/10.1111/nph.14219, http://onlinelibrary.wiley.com/journal/10.1111/(ISSN)1469-8137.

Swain, N., Mishra, S., 2019. A review on the recovery and separation of rare earths and transition metals from secondary resources. Journal of Cleaner Production 220, 884–898. Available at https://doi.org/10.1016/j.jclepro.2019.02.094.

Tao, Y., Shen, L., Feng, C., Yang, R., Qu, J., Ju, H., Zhang, Y., 2022. Distribution of rare earth elements (REEs) and their roles in plant growth: a review. Environmental Pollution 298, 118540. Available at https://doi.org/10.1016/j.envpol.2021.118540.

US Geological Survey. 2022. 2022 Final list of critical minerals, FR Doc. 2022-04027. URL: https://www.federalregister.gov/documents/2022/02/24/2022-04027/2022-final-list-of-critical-minerals [accessed: 22.05.2025].

Vaculík, M., Mišljenović, T., Lukačová, Z., Jakovljević, K., Podar, D., Kováč, J., 2024. Zinc hyperaccumulation in plants: mechanisms and principles. Zinc in Plants: Current Knowledge and Recent Advances. Elsevier, Slovakia, pp. 1–39. Available at https://doi.org/10.1016/B978-0-323-91314-0.00017-X, https://www.sciencedirect.com/book/9780323913140.

van der Ent, E., 2023. Mining with plants, Chenmistry & Industry 87 (11), 30–33. Available at https://doi.org/10.1002/cind.10190.

van der Ent, A., Baker, A.J.M., Echevarria, G., Simonnot, M.-O., Morel, J.L., 2021. Agromining: Farming for Metals: Extracting Unconventional Resources Using Plants Mineral Resource Reviews. Springer International Publishing. Available at 10.1007/978-3-030-58904-2.

van der Ent, A., Baker, A.J.M., Reeves, R.D., Chaney, R.L., Anderson, C.W.N., Meech, J.A., et al., 2015. Agromining: farming for metals in the future? Environmental Science and Technology 49 (8), 4773–4780. Available at https://doi.org/10.1021/es506031u, http://www.pubs.acs.org/journal/esthag.

Vaněk, A., Chrastný, V., Mihaljevič, M., Drahota, P., Grygar, T., Komárek, M., 2009. Lithogenic thallium behavior in soils with different land use. Journal of Geochemical Exploration 102 (1), 7–12. Available at https://doi.org/10.1016/j.gexplo.2008.10.004.

Vaughan, J., Riggio, J., Chen, J., Peng, H., Harris, H.H., van der Ent, A., 2017. Characterisation and hydrometallurgical processing of nickel from tropical agromined bio-ore. Hydrometallurgy 169, 346–355. Available at https://doi.org/10.1016/j.hydromet.2017.01.012.

Wan, X., Zeng, W., Lei, M., Chen, T., Li, Y., 2024. Ex-situ phytoextraction using *Pteris vittata* as a sustainable way to protect soil resource. Environmental Technology & Innovation 34, 103638. Available at https://doi.org/10.1016/j.eti.2024.103638.

Wang, D., Liu, H., Ma, Y., Qu, J., Guan, J., Lu, N., et al., 2016. Recycling of hyper-accumulator: synthesis of ZnO nanoparticles and photocatalytic degradation for dichlorophenol. Journal of Alloys and Compounds 680, 500–505. Available at https://doi.org/10.1016/j.jallcom.2016.04.100.

Xiao, H., Zhang, Z., Li, F., Chai, Z., 2003. Study on contents and distribution characteristics of REE in fern by NAA. He Jishu/Nuclear Techniques 26 (6), 420. Available at http://www.hjs.periodicals.net.cn/default.html.

Xiao, T., Guha, J., Boyle, D., Liu, C.Q., Chen, J., 2004. Environmental concerns related to high thallium levels in soils and thallium uptake by plants in southwest Guizhou, China. Science of the Total Environment 318 (1–3), 223–244. Available at https://doi.org/10.1016/S0048-9697(03)00448-0, https://www.elsevier.com/locate/scitotenv.

Xie, C., Xiao, Y., He, C., Liu, W.-S., Tang, Y.-T., Wang, S., et al., 2023. Selective recovery of rare earth elements and value-added chemicals from the *Dicranopteris linearis* bio-ore produced by agromining using green fractionation. Journal of Hazardous Materials 443 (B), 130253. Available at https://doi.org/10.1016/j.jhazmat.2022.130253.

Xu, W., Xiang, P., Liu, X., Ma, L.Q., 2020. Closely-related species of hyperaccumulating plants and their ability in accumulation of As, Cd, Cu, Mn, Ni, Pb and Zn. Chemosphere 251, 126334. Available at https://doi.org/10.1016/j.chemosphere.2020.126334.

Xue, S.G., Chen, Y.X., Reeves, R.D., Baker, A.J.M., Lin, Q., Fernando, D.R., 2004. Manganese uptake and accumulation by the hyperaccumulator plant *Phytolacca acinosa*

Roxb. (Phytolaccaceae). Environmental Pollution 131 (3), 393–399. Available at https://doi.org/10.1016/j.envpol.2004.03.011.

Yan, T., Wang, X., Liu, D., Chi, Q., Zhou, J., Xu, S., et al., 2021. Continental-scale spatial distribution of chromium (Cr) in China and its relationship with ultramafic-mafic rocks and ophiolitic chromite deposit. Applied Geochemistry 126, 104896. Available at https://doi.org/10.1016/j.apgeochem.2021.104896.

Yan, X.L., Chen, T.B., Liao, X.Y., Huang, Z.C., Pan, J.R., Hu, T.D., et al., 2008. Arsenic transformation and volatilization during incineration of the hyperaccumulator Pteris vittata L. Environmental Science and Technology 42 (5), 1479–1484. Available at 10.1021/es0717459.

Zhang, X.H., Liu, J., Huang, H.T., Chen, J., Zhu, Y.N., Wang, D.Q., 2007. Chromium accumulation by the hyperaccumulator plant *Leersia hexandra* Swartz. Chemosphere 67 (6), 1138–1143. Available at https://doi.org/10.1016/j.chemosphere.2006.11.014.

Zhang, J., Wu, S., Xu, J., Liang, P., Wang, M., Naidu, R., et al., 2021. Comparison of ashing and pyrolysis treatment on cadmium/zinc hyperaccumulator plant: effects on bioavailability and metal speciation in solid residues and risk assessment. Environmental Pollution 272. Available at https://doi.org/10.1016/j.envpol.2020.116039, https://www.journals.elsevier.com/environmental-pollution.

Zhang, X., Houzelot, V., Bani, A., Morel, J.L., Echevarria, G., Simonnot, M.O., 2014. Selection and combustion of Ni-hyperaccumulators for the phytomining process. International Journal of Phytoremediation 16 (10), 1058–1072. Available at 10.1080/15226514.2013.810585.

Zhang, X., Laubie, B., Houzelot, V., Plasari, E., Echevarria, G., Simonnot, M.-O., 2016. Increasing purity of ammonium nickel sulfate hexahydrate and production sustainability in a nickel phytomining process. Chemical Engineering Research and Design 106, 26–32. Available at https://doi.org/10.1016/j.cherd.2015.12.009.

Zhang, R., Zhang, Y., Goei, R., Oh, W.-D., Zhang, Z., He, C., 2023. Sustainable utilization of *Sedum plumbizincicola* as superior hydrochar for efficient nutrients recovery. Journal of Environmental Management 344, 118441. Available at https://doi.org/10.1016/j.jenvman.2023.118441.

Zhong, D., Zhong, Z., Wu, L., Ding, K., Luo, Y., Christie, P., 2016. Pyrolysis of Sedum plumbizincicola, a zinc and cadmium hyperaccumulator: pyrolysis kinetics, heavy metal behaviour and bio-oil production. Clean Technologies and Environmental Policy 18 (7), 2315–2323. Available at https://doi.org/10.1007/s10098-016-1150-y.

Zhong, D.X., Zhong, Z.P., Wu, L.H., Xue, H., Song, Z.W., Luo, Y.M., 2015. Thermal characteristics and fate of heavy metals during thermal treatment of *Sedum plumbizincicola*, a zinc and cadmium hyperaccumulator. Fuel Processing Technology 131, 125–132.https://doi.org/10.1016/j.fuproc.2014.11.022.

Index

Note: Page numbers followed by "*f*" and "*t*" refer to figures and tables, respectively.

R

S

T

Printed and bound by CPI Group (UK) Ltd, Croydon, CR0 4YY

14/09/2026

02216409-0002